Space Rangers

Copyright 2014 Lordship Investment Corporation. All rights reserved. No part of this book may be reproduced, stored in a retrieval system or transmitted by any means electronic, mechanical, photocopying, recording or otherwise, without the written permission from the author.

Introduction

I spent my childhood reading and collecting comic books. I loved reading about the heroes of Marvel and DC Comics.

Get ready for an adventure. We are going to delve into the world of the Space Ranger. While we are looking at it we will see the romance develop between Matthew and Blair.

The year is 2425. The planets in our solar system have been under attack by a collection of alien races bent on conquering our planets. We have united with other galaxies to form GAP, the Global Association of Planets. Together we formed "Space Rangers", interplanetary police officers who keep the peace in our galaxies.

Matthew and Blair met while on earth. They both signed up and became partners in the GAP program. Through the course of this journey they fall in love.

This is the first of 3 books showcasing the exploits of the Space Rangers. Look for the other books as they become available to complete your collection.

I hope you enjoy this journey through the eyes of 'Space Rangers".

Dr. Jeff Davis

Chapter 1 - Matthew

The year was 2425. For the last 150 years there was an ongoing war in the galaxies. This war had brought havoc to many regions of space. It was between what some consider the peaceful races and the warring races.

Planets that had the technology began to look to space as the "next frontier." They would explore the planets in their own galaxy. If the planet was habitable they would seek to inhabit them. These newly inhabited planets soon had thriving colonies on them. Some of these planets were rich in minerals and metals which made them a goldmine for the host planet. The colonies and the host planets began to thrive.

There were some planets that did not have ample resources on their sister planets. So these planets began to attack the other planets that had these resources. It forced the parent planets to defend their colonies.

The worse of these attackers was a group called DASHOOOR. They were a collection of thieves and mercenaries who were relentless in their conquest of the planets with resources. Many of these galaxies folded to DASHOOOR. It had gotten so bad that they began to erect prison planets to house the people they conquered. They plundered their planets for all available resources and force the people into slavery. They were very cruel taskmasters and many died at their hands.

The leaders were from a race that had reptilian blood which made them very cold blooded. They were highly intelligent and felt that all the other beings in the galaxy was beneath them. Given their size, strength and intelligence, they were formidable enemies indeed. The average alien who went up against them in combat died. They clearly had the advantage. On average they were tall and as strong as 10 men.

Even being out manned and outgunned there were races that were not willing to give in to the tyranny. For those who wanted a chance to fight back they formed an allegiance, the Galaxy Alliance of Planets. This group not only protected the planets in their systems but they began to police the galaxy and help other planets stand against the enemy. The GAP began to grow in membership and military power as the joining planets added their armies to the fold. Soon the balance of power was shifting as GAP began to slowly take ground.

Earth formed an alliance to represent the military and sent representatives to join GAP. This meant joining in the ongoing war against DASHOOOR. The GAP committee voted to place an outpost outside the Milky Way Galaxy making it under the protection of the Earth alliance. There was a lot going on in space and many of the inhabitants of the earth had no idea of how dangerous the galaxy had become.

This is the story of how the Galaxy Rangers came into being. This group was the premier unit of the Galaxy Alliance. It was made up of specially trained soldiers who needed to have an edge to compete with these alien invaders. As we look back we can trace the origins of the first Galaxy Rangers who took the fight to the Dashooor community.

It was a cold winter day in the city of Chicago, Illinois in 2401. Edward was just coming in from work which was only a few blocks away but he came home buried in snow. He walked through the door and shook the snow off. His wife Patricia came to meet him and kissed him at the door. "How was your day honey" she asked. "Life at the factory never changes. We build things, pack things and ship things. Then we go home."

Edward had migrated to the Midwest from down south. He figured he could come to Illinois and find a job much easier than he could in his hometown of Alabama. He had no particular set of skills but he was a hard worker, fast learner and made it a point to give whatever activity he was involved in 100% of his attention.

He met his wife Patricia in Illinois. She was originally from the city of San Francisco, California and moved to Illinois after going through a painful divorce where her former husband ended up with their 2 children because he had a more stable job and could provide a better life. Broken in spirit Patricia moved across the country until she ended up in Illinois. She met Edward at a local grocery store. Here was a man who grew up on a farm and understood fruits and vegetables like no one she had ever seen. He was examining each piece of fruit with an eye for detail, ripeness and apparently making himself a meal to be cooked later that day. They struck up a conversation and found that they both like fresh fruits and vegetables with home cooked meals.

After months of dating they were married. Soon after their family began to grow with the addition of children, a son (Mitchell) then a daughter (Miriam). Their last child was a boy by the name of Matthew. Each child was given a first and middle name to go along with their last name of Lincoln. So this son became Matthew Abraham Lincoln. To say that his name would be the subject of some discussion was an understatement.

This was their life and it was a good life. They lived in a cozy bedroom apartment and had a home filled with love. Edward went to work and was home each weekend. Pat made sure the children were well provided for. She loved to walk through the house singing and encouraging the children to join her. Although no one in the family had much of a voice they tried very hard anyway.

One night Pat spoke to Edward about their apartment. Although it was nice and cozy it was really too small for 5 people. "Edward, we should seriously look for a house. We need a place where the kids will have a bedroom and we can have more space. I would enjoy a larger kitchen to cook our meals in."

"That would be very nice Pat but you know I work all day. I can't go looking for a house. And we don't even have enough money to buy a house. Let's just be content with a roof over our heads."

This was how Edward was. He wanted things to stay the same because he hated change. And even though there were some things they needed to improve their family life he was reluctant to make the changes. As long as there were no real problems, he was okay.

Pat was different. She was a dreamer. She was the reason they were in the apartment now because she found it when they got married so that they could have a place of their own. She knew that in order for her family to keep making progress they needed a larger space.

"Honey, I can look for a house in the area so we won't have to move too far. And I will get us approved for a loan. If I can do all that will you agree to look at the house and if you like it, we can move." Edward agreed thinking to himself she would never pull this off. He was wrong.

By December of that year they were moving into their new home. It was 3 stories high, plenty of space and everything that was hoped for. The children had the best Christmas ever. Family and friends came by to celebrate the new home. It was wonderful.

As the children got older and went to school Pat began to notice things about Matthew. He seemed to pick things up very quickly. His kindergarten teacher told Matthew he would never amount to anything because he was always getting in trouble. Pat knew that when he was bored, Matthew made things up to do. She told the Principal the school needed to challenge him more. And she told Matthew "No one's opinion of you is more important than your opinion of yourself. Refuse the negativity and reach for the stars.' Matthew took that advice as a five year old child and graduated grammar school at the top of his class.

It was during these early years that Matthew discovered he loved Astronomy. His family got him a telescope for Christmas one year and he spent countless hours looking at the stars. He got books and read about space. He loved watching old reruns of space movies.

Matthew fell in love at 14. He met the love of his life, Candace. She was a cutie pie and he smiled from ear to ear when he spoke to her. He loved the phone calls and the way she smiled at his jokes. It was really nice to have someone who made him feel so special.

Pat was not feeling the relationship. She felt her son had fallen way too hard for a girl and he was just a baby (her baby). Since their family lived a block away Pat made her way down to Candace's house and met her family. Candace's mom was a single parent with 3 teenage daughters. To say they were fast and undisciplined would be an understatement. These girls were both pretty and promiscuous and this bothered Pat. She did not want her son to grow up too fast or to experience things in life that should wait until their time.

One day Matthew and Candace were sitting outside and Candace reached over and kissed Matt. This was his first kiss. His first time touching a girl on the hips and letting his hands run down to her butt. He reached back real fast and apologized. "I am so sorry. I didn't mean to touch you like that. Won't happen again".

"Why not? You are my boyfriend, right? It's okay for you to touch me." So touch is what they did. Well into the night.

Being so young and being exposed to feelings at this age was not good for Matthew. What it did was awaken feelings inside of him that could ultimately lead him to doing something he was not ready for. Even though he was not ready for what was sure to come, his mother was. And she was not about to let her boy get caught up in situations he was clearly too young to deal with. She spoke to him the next morning. "Matt, I know you are crazy about that girl down the street. Where do you see the relationship going in 6 months or a year?"

"I don't know mom. I want to take it one day at a time. Its summer time and right now we just hang out. We are not doing anything you have to worry about."

"Ummm hmmm" she said to him with her head leaning to one side. "Son, I have been where you are. My first boyfriend was the best, a real piece of work. I ended up marrying him and found out he was the greatest jerk of all. I don't want to see you end up like that.

"I won't mama. I promise." Then Matt got up and headed for the door.

"Where are you going?" Pat asked.

"No where. Just going to play baseball. Why"?

"Because I need you to do something before you leave."

"Sure mom. What you need"?

"I need you to break up with your girlfriend. You are too young for a serious relationship. I want your focus on books and learning, not girls."

"But mom, we are not doing anything wrong. We just hang out, talk and have fun. Why can't I see her anymore?"

"Son, we are not going to have a discussion about this today because I know you won't understand. What I know you do understand is when I tell you to do something, you can do it."

Matthew was so angry all he could do was burn up inside. He didn't say a word; just dropped his head and walked out of the room.

For the next 2 days it was deathly silent in the Lincoln home. Matthew refused to look at his mother and she sensed his coldness. But she also knew it was the right decision and Matthew had learned how to follow directions even when he did not understand.

The break up was spiteful. Once he went and told his girlfriend they could no longer see each other she threw a tantrum. She called him all out of his name, cursed him and told him she didn't want to see him again. The sad thing was that while she was standing in the doorway throwing a fit there was another guy in the apartment a few feet behind her. He told her to "close the door, drop that loser and get back there so they could get busy." Matthew stood in awe. He had chosen to be angry over the split up with his mother and although he was hurting his girlfriend had already moved on and was kicking it with an older guy. Matthew thought to himself "what a fool I have been" and walked away.

When he got home and saw his mother in the doorway he put his head down to walk away. Then he stopped. He looked at his mother and ran to her giving her a hug. "Mom, I am sorry. I should not be mad at you. Please forgive me."

Pat smiled with a tear coming down her eye. She said "Matt, I know what I said hurt but let me tell you why I said that. I know your girlfriend has a couple of sisters and I know their mother.

Those girls are fast and I figured their baby daughter was no different. I do not want you caught up in someone's mess."

"I hear you mom. I hear you."

"Just remember son. Other will love you, some a lot and some a lot less. But no one on this planet will ever love you more than me. Remember that as you date."

Matt held his mom for a long time.

As time went on Matt dated a lot of different girls. He was a good student and gifted athlete who loved wrestling. He was on his school wrestling team along with the track and chess teams. He began to make a lot of friends and had a great social life.

That is when he met her; Larita Reynolds. This girl was hot. She had short black curly hair, a great shape and the cutest smile. Matt wasn't sure she would go for him since so many of the other guys were trying to get next to her. But for some reason she was attracted to Matt. And he was attracted to her.

They would go out to lunch together and they were in the same math class. Matt never asked her to be his girlfriend but pretty soon it was assumed since the two of them were always together.

She would go to his basketball games, track meets and chess tournaments. One day Larita asked Matt a question while he was getting his stuff after a chess tournament. "Matty (her nick name for him), what do you think will happen to us once we graduate?"

"Hmmm, I had not thought much of that. Figured we would continue this great relationship we have. You would like that wouldn't you?"

"Yes" she said, walking slowly up to him and standing so close he could feel her breathing on him. "I want to be where you are. Not just here in school, but always."

"What are you saying, Larita?"

"I am saying I want to be with you for a long time. I love you."

It was the first time the word had been used in their relationship and it took Matthew back. He stood there silent for a minute then he answered; "I am surprised to hear you say that. I knew you liked me but love is a very strong word."

"So you don't love me?"

"I, I ……… I don't know if I even know what love is. Sorry."

Larita put her head down and left the gymnasium. Matthew put his head down because he knew she deserved a better answer than that.

Walking to his car Matt saw her sitting alone. He walked over to her, put his arm around her and then lifted up her head with his hands so that she could see his eyes.

"Baby I am sorry. You are so good to me and I am afraid to tell you how I feel."

"Why?"

"Because the last time I put my heart out there, it was crushed. I have been very guarded about my emotions ever since."

"I understand. Well I won't crush your heart. I promise to hold it next to mine and treat it like gold. You are precious to me and you make me feel special when I am around you."

"Thank you so much. I really needed to hear that." Then he leaned forward and kissed her passionately.

Larita could feel herself melt into his lips as she cried tears of joy that this man was her man.

He told Larita his plans and she agreed that no matter which direction he went in, she would be there for him. Matt felt blessed to have such a dependable girl by his side. And she was cute to boot.

Matthew looked forward to the day he would leave home and join the service. His grades were good in school but Matthew felt he would learn leadership better by serving in the armed forces. His goal was to one day become an officer and lead men and women.

His family was not all that happy with his decision. His father tried to talk him out of it and moving on to college to get his degree and get a good job. Edward was a good father but he lived his life on the safe side. He felt that if you had a job you should be happy. He struggled with Matthew's decision to sign up for the GAP.

Pat was a different story. She was only concerned about Matthew's safety. "If you agree that you will not be the first one to run on the line and risk his life, I will support your decision. I will talk to your father. I do not want to lose my son. Can you understand that"?

"Yes mom, I will not run and offer my life recklessly. I want to lead people back to their families, not to die in the battlefield" said Matthew.

This was the focus for Matthew. He wanted to explore his options but he could not get away from the feeling that he wanted to be a leader in the military. This desire would be appropriate for the one who would one day lead the first GAP unit into space exploration.

Matt was torn about what to do. He had wanted to go to college and had the grade to get into a good school. The only problem was he was not sure as to what to major in. He wanted to join The Galaxy Association of Planets (GAP), the military which watched over the galaxies and provided law enforcement. But if he were to get an assignment it would mean the end of his relationship with Larita. He would be out at a space station and would not see her for months at a time. He felt that was unfair to such a good woman and she should be free to find someone who she could be with while still on earth.

He called her up and asked her to meet him at the park they went to often. She agreed to be there in an hour.

"Larita, looking as good as usual. Glad you could come out and see me."

Larita gave a little smile but Matt could tell something was wrong. They walked over to a bench. He waited until she started the conversation.

"Matt, I have something to say to you. You know I love you and I want you to follow your dreams. But as I was talking to my family they are not too keen on us trying to have a relationship if you go off to GAP. You may be gone up to a year at a time. I would miss you so and I just ….."

Matt reached over and put his finger on her lips. "I understand. I was thinking the same thing. I want you happy. You need to find someone you can share your life with and do those things you need and want to do. I believe I am supposed to work for GAP. It's my destiny. I don't think I can walk away from it."

"No, don't you dare. I am not here to stand in your way. I am here to say goodbye. I love you Matt."

"I love you too. Good bye."

As she got up and left Matt felt his heart sink. He just let an incredible woman get away from him. And for what? To fly around in space? He knew he made the right decision. It just hurt to watch her walk away.

Space travel was at an all-time high. And with it were incidents of crime in space. The Galaxy Association of Planets recruited a team to patrol the airways and enforce planetary law. This would be a high honor but it would mean leaving earth and working with a team of soldiers who were referred to as Space Rangers, intergalactic police officers.

After careful consideration Matt made his way down to the Recruiters office. He listened to the presentation and learned how he could still go to college and earn his degree as the GAP had their own academy for cadets. This way Matt felt he could kill two birds with one stone. The possibilities of what he could major in were great so he knew he would not be settling. He actually had his heart set on studying Astronomy.

He went into a room with over 100 eager candidates. Matt did not know what to expect. He received information to read about joining the armed forces and was hiding his excitement as he sat in his seat. He saw with men and women, all under 30 years old. Apparently GAP was focusing on reaching a young group of recruits who showed promise academically and athletically.

A test was administered which was unlike any tests Matt had taken before. Instead of asking him a lot of questions this test was more like a video game. In it he was given 5 people on his team and their objective was to rescue a captured diplomat. At first he thought the objective was to go and fight his way until he could get to the diplomat. But the game was designed to see if he could lead his team to make the rescue. So he programmed each team member to do a part of the rescue and implemented his plan. As he completed the test Matt was curious because the test did not reveal your score. It only said "game over."

As Matt took his seat he saw three officers walked in, dressed in their various uniforms. The officers surveyed the room as they headed up to the platform. One of them began the presentation;

"Thank you for coming out today. We are so excited about the opportunity you have to join us in keeping our world safe. We are always looking for the right people to join our team."

"You realize that we are looking for the best to join GAP. We still have the Earth military in place to protect our countries but each nation knows the real threat is in space. This is a good time in that we have united our armed forces on Earth and each nation sends it's best to the GAP academy. For those who make the cut you will be working in space with soldiers from all over the world. The best of you will be eligible to become a Space Ranger."

"Our space station is responsible for protecting our airspace. GAP has an outreach base located near the end of our universe to protect us for alien invasion. We have the ability to scan neighboring planets and work hard to keep our space ways safe. The Space Rangers patrol the neighboring galaxies and enforce intergalactic laws. They are empowered by the GAP council and can arrest, capture and detain criminals. It is a great opportunity to extend needed protection for the human race." The officer was passionate in his presentation and opened the floor up for questions.

"What must we do if we are interested in joining GAP?" said one hopeful participant.

"We will talk to you individually. We have access to your background info and your grades so if you are chosen to be considered for enlisting you will be approached by one of my fellow officers."

"Oh we cannot just choose to be a part of GAP?"

"No, you must be chosen. We have thousands of applicants from all over the world."

Matt waited until the meeting was over and hoped he would be called. He seemed to be by himself and when no one approached him he thought he had not made the grade. He put his head down and headed for the door. As he got up he heard his name called from the front;

"Matthew Abraham Lincoln. Could you please make your way up to the front?"

Matt walked to the front and the recruiter took him to a side room to talk to him.

"Hi Matt, I am Sargent Paul Simon, head recruiter for GAP academy. I want to tell you congratulations. We have reviewed your school scores, background record and grade from the simulation you did today. Based on these results I am prepared to offer you a chance to become a member of GAP academy."

"Wow that is great. I didn't know you had all that info on me."

"We do our homework Matt. We look for the cream of the crop. Tens of thousands of people apply each year but we can only take the best from the world to be a part of the academy. And the best of that group joins GAP. You are given a chance to do something great. Now I need to know if you want to check out this opportunity further.

"Yes, yes I do."

"Then there is a lot of work to be done before you leave for GAP academy. What I need you to do is go over this application, sign it and say you want to be enlisted. Then it begins."

Based on the conversation and the presentation, Matt enlisted. He could not wait to go home and tell his parents of his decision.

His father was glad. He believed that the DASHOOOR were a ruthless bunch of lizards who needed to be put down hard. He knew people who were devastated by them and was ready for Earth to get into the battle and kick some butt. His mother was glad that Matt was happy but still afraid she would lose her son.

Matt's departure date was only 3 weeks away and he had much to do. There was a lot of paperwork to be filled out along with videos to watch outlining the history of GAP. He also was fitted for his training camp uniform and equipment. They told him to go home and begin to do exercises to get his body ready for the grueling experience he was about to have. Of course he had to get a haircut to look more like a soldier.

When it was time for him to finally depart for training camp Matt looked like a soldier. He had been swimming almost every day, eating right and exercising. He felt good and strong. When he did get dressed in his training outfit everyone who saw him had to say how good he looked.

'Wow Matt. You look like a GAP officer. Way to go man" were words from his Uncle Billy who came to say goodbye. Billy was the family sports expert and was coaching Matt on how to excel in whatever sport he chose to play since Matt was a child.

"Uncle Billy, thank you for being in my life. I won't forget how you prepared me for this day." With that Matt hugged his uncle. "I am expecting some great things from you Matthew. You have natural abilities which is a blessing. But more importantly you have a head on your shoulders. You have shown what it takes to make a difference in the world. You shall go far young man."

Just then Matthew heard a very family voice coming into the room. "Matty, you are such a brat. But you know I love you anyway" words spoken by Lucy, Matt's cousin and classmate. Since they were the same age she spent many years in the same classroom as Matt.

"Lucy, I am gonna miss you sitting in class with me. But I will be back" and with that he hugged her. She began to tear up and kept repeating "I am not going to cry. I am not going to cry."

His siblings came up and offered congratulations. His dad hugged him hard and said "I am so proud of you. Go and help make this galaxy safe again."

When he got to the door he saw his mother. She had tears in her eyes. "I am sad to see you go, proud of your decision, scared for your safety and thankful that you are such a smart kid I know you will take care of yourself." With that she hugged him, kissed him and said "Come back to us in one piece. You promised."

"I will mom. I promise."

The transport ship was hovering above the grass. Lights began to flash and they called Matt to come out. As he grabbed his stuff Matt looked at his family one last time. It was hard to believe that he would be going away from them all. They had been the most supportive family he could have asked for. A tear went down his cheek, he waved goodbye and ran to the transport.

Once he was on board he saw a ship full of new recruits. Some were happy and giddy. Some, like him, were very sad. Matt went and sat down and just thought about what this decision he had made would affect those people he left behind. He was scared but excited at the same time. And the last thing he wanted to do was have them worry about him.

Matt looked out the windows and saw the earth getting smaller and smaller. He knew that eventually he would be taking off for space and it would be a long time before he would see his planet Earth again. It was a beautiful planet. And he was going to work to make sure it stayed that way.

Chapter 2 – Blair

Arkansas is a small family friendly state. Growing up in Little Rock meant going to a local school, having friends all along your block and going fishing. This was the community Mitch and Susan Hutchinson decided to raise their children in. They were both from the South and loved being there. The sense of community was intense.

Mitch was an auto mechanic who worked 12 hour days. His wife Susan taught in the local school. To their family was born a precious little girl, Blair. She was the light of both of their lives and the reason for much of the joy they had in their household each day. Even after long days at work they both looked forward to coming home and having a quiet dinner together no matter what time it was.

Mitch had a brother who was a member of the GAP. He was a career officer and would come to visit Mitch in Little Rock whenever he was in town. Susan had two older brothers who also were in the military so when all the family came together Mitch would catch it for not being in the military like the rest of them.

Blair was a very gifted athlete. From a young age she showed talent in sports, especially running and volleyball. Her parents went to her events and showed their support for this daughter of theirs who was on her way to doing great things with her life. She was on her way to being offered scholarships for college.

Blair met James, another athlete at her school who took an immediate liking to her. He was tall, muscular and a real looker. Blair knew that just having a conversation with him made her heart skip a beat. She worked really hard to conceal how excited she was in his presence. He knew that and played on her obvious affection of him.

Movies. Dinner. His good night kiss was the first kiss Blair had ever had with a boy. And James took advantage of that. He gave her a peck on the lips as she closed her eyes. Then while she was basking in the feeling of that first kiss James proceeded to put his tongue in her mouth. She was taken aback but then she realized this felt kind of nice. So instead of resisting she went with the kiss. And it was a long kiss indeed.

James took this as a sign that she wanted more from him so as he continued to kiss her he let his hands wander. First they went up and down her back, and then slowly he let them rest on her hips. He didn't move but just kept holding her and kissing her. She was feeling things she had never felt before.

When the kiss finally ended James let his hands wander to touch her behind to see if she would say something. Blair did notice the hand on the forbidden zone (as her uncle's called it) but she said nothing. She was so enamored by this great guy James who was paying attention to her that nothing else mattered.

"Susan, we ran your tests and ran them again to make sure we had not missed anything. I am sorry to tell you that we found traces of cancer in your bloodstream."

Mitch's lip almost hit the floor. He was stunned to believe that his health conscious wife, the woman who exercised more than him and took such good care of herself could be diagnosed with cancer. Susan looked in shock and put her head down, beginning to cry. Mitch reached over and held her hand, saying not a word.

"We have a regiment developed to address your type of cancer. I must tell you it is a very aggressive form you have and we must begin treatments immediately. We will start with a full work-up and then place you on chemotherapy. Afterwards we want to go in and do surgery to get rid of as much of it as we can, hopefully all of it. Then we finish with radiation therapy."

"I must ask that you both get incredibly focused on fighting this thing. Let your loved ones know. Susan you will get discouraged but I believe you have great support to make a real fight of this. I want you to know me and my staff will be with you and your family every step of the way. We will not let you face this alone" said Doctor Reynolds. Mitch and Susan could feel the sincerity in his voice and this provided comfort.

As they left the doctor's office holding hands, they said nothing. Then they looked each other in the eye and one word came out of both of their lips at the same time; "Blair".

How do you go home and share this type of news with your children? Your loved ones? They had no clue. But they knew that Blair deserved to know and would be more hurt if they didn't share it.

They got home and made dinner as usual. When Blair came in all excited about getting an A on her quiz she couldn't wait to share the news. She went on and on about school while her parents listened quietly. Soon it dawned on Blair that something was wrong and she began to ask questions.

"Okay guys, what's up? You haven't said two words between you since I got home. Is there anything you wish to tell me?"

"Maybe you need to come into the living room so we can all sit down and have a family chat" said Mitch. Blair knew at that point something was wrong. Very wrong.

They all sat and Mitch began. "Honey, me and your mom went to the doctor and we have some news to share with you. Your mother was diagnosed with cancer. As you can tell we are still in shock over the news."

Blair turned and looked at her mother and began to tear up. "Oh no, not mom. Not my mom. How did this happen?"

Susan looked at her daughter with tears in her eyes and said "Baby we don't know. The doctor said it just happens. Not because we were foolish about our health. It's no one's fault. It just happened."

Blair got up and ran to where her mother was sitting and put her head on her lap and cried. Mitch got up and walked over to these two women in his life and wrapped his arms around them both. They all sat and cried for a long time. Blair was close to her mom because she was her best friend. Throughout her life her mother was always there. Never judging, just accepting.

Drying his eyes Mitch spoke up. "Look, we are a family of fighters and this is no different. We are going to learn all we can about this cancer go through whatever the doctors deem we should and remain positive. I know it hurts. So let's cry now. Afterwards, let's get ready to go to war."

Mitch and Blair were super supportive to Susan. Mitch cut back on his hours at work to be at home more. Blair came straight from school and assumed her mom's household chores. They ate dinner together each night and laughed when they could. They told their relatives and all the uncles and brothers came by on a regular basis to lend a hand. Their wives began bringing food so the family wouldn't have to worry about cooking so much.

Susan began her regiment of treatments and in the beginning she was strong. But as days turned into weeks she got weaker and weaker. She fought hard to stay encouraged but it was a battle. Soon she was having trouble getting around. Mitch and Blair tried as hard as they could to keep up a good front but behind closed doors, with no one looking, they both cried.

One afternoon Blair came home. It was her turn to look after her mother since her dad was working late. It was all he could do to get up and go to work because he never wanted to leave his wife's side. But she insisted and Blair promised to be there at home.

Susan was sitting up in her chair in the living room when Blair came in. 'Hi mom."

"Hi honey. How was school?"

"Fine. How was your day."

"It's been a good day baby. I am so blessed."

"Good to hear mom. I am making me a snack. You want one?"
"No. But when you are done, come and sit by me. I want to talk to you."

Blair got her sandwich and came into the living room with her mouth full. She sat down on the couch and looked at her mom."

"You look good today mom."

"Thanks baby. I want to talk to you about your future and your dad."

Blair sat up and put her food down. "Okay mom, what's up?"

"Honey, these are some trying times our family is going through. But I want you to know that I am committed to being honest with you every step of the way."

"Okay. I like that."

"You are almost 18 and getting ready to graduate. You have your whole life in front of you and I believe an awesome future. My biggest regret is that I am not going to be there with you."

"Oh mom. Don't say that."

"It's true. I have been fighting for months and you and your father are awesome." Susan began to tear up. "God could not have put me in a better family with such great people. I cannot tell you how much I have grown to love you both so much as I have watched you care for me, work together and cry when you think I am not looking. I feel your pain and I am so sorry you have to go through this."

Blair began to cry. "Mom, please. It's been hard, yes it has. But love is there for the hard times. You would do no less for me or dad. We love you too."

"Susan put her head down and looked back up, smiling. " You see what I mean? Even when you hurt, you still support. You are going to be an awesome wife to some young man someday."

"Oh mom!" Blair began to blush.

"Listen to me. I never want you to believe that you are not special or of no value. Anyone who ends up with you will have a treasure in his hands. I want you to promise me to stay true to yourself, follow your heart, pursue your dreams and make a life you not only enjoy but will be proud of."

"I will."

Susan reached over and put her palm on Blair's face. "The greatest thing I have ever done is give birth to you. The greatest joy I have ever felt is being your mother. And the greatest love I could ever have is for this family. Stay strong and help your father. He is the best man any woman could ask for. But I know that he will need help getting through the loss of me. Don't be too hard on him. Once he gets his bearings, he is a force to be reckoned with."

"I will mom. I promise I will." Blair got up and hugged her mother. They hugged for a long time.

"Now put me down for a nap dear. And call your father. Tell him I want him to come home early tonight so we can talk."

Mitch got the message at work and came home as fast as he could. When he went in Blair was in the kitchen cleaning up.

"Hi hon. How's mom?"

"She is good dad. Just resting. She wants to talk to you."

"Okay." He dropped his lunch bag on the table and went into the bedroom.

"Susan, how are you?"

"Fine honey. Just fine. Thanks for coming home so quickly."

"Of course. Is anything wrong?'

'No it is not. I just wanted to see your face and tell you that I love you."

Mitch leaned down on the side of the bed and put his hands in hers. "You know I love you too."

"Tonight will be my last night at home. I need you to take me to the hospital."

"Why? Is something wrong?'

'No. But I want to have the doctors look at me tonight. Can you and Blair do that for me?

"Yes." Mitch called for Blair and they rushed to get Susan to the hospital.

As they sat in the waiting area the doctor came out to meet them.

"Mitch, Susan has gone into a coma. I don't think she will be with us much longer. I think it best if you and Blair went in and said your goodbyes."

They both entered the room and stared at this amazing woman. Getting on both sides of the bed they said goodbye. Then they sat in the room and just sat there.

Susan fought hard. But in the end she lost. And Blair's life was never going to be the same again. She lost not only her mother but her best friend. Most of her relatives were male so her mom was the foundation. She felt truly alone.

The community came out in the masses to support the family. Susan was the rock that bound together many families. She was loved a lot. Her influence was already missed when she took sick and had to back off on things she use to do. She was confined to hospital visits and stayed at home. With her death a real void was created in this small, close knit community.

After the funeral Mitch began to slowly lose himself. He lost his joy and his will to live. He slipped deeper and deeper into depression. Blair could no longer reach her dad who seemed unhappy every day. She was fighting her own battle with being discouraged. Now she had to find a way to encourage some else. But who was there for her? Who was lifting up this young 18 year old woman?

She had very few friends to share her life struggles with. So she began to internalize her feelings. Instead of being open and care free, she was now more reserved and private. The death of Susan drove both Mitch and Blair to places neither had ever been at before. They did not know how to deal with their loss and instead of getting closer, which they both needed, they grew farther apart.

Mitch started to drink. Blair started to gravitate to boys at school. Stan and Andy, Mitch's brothers, tried real hard to reach him. But he fought them tooth and nail. Mitch had a habit of going inside instead of looking outside for support. Mitch was bitter and angry with himself for not being able to save the love of his life.

"Look bro, we know you hurting. We loved her too. She was a great wife, great mother and the closest person to a sister we ever had. We hurt with you. Don't shut us out. Close us in with you and let's cry together. You are not alone" said Stan.

"I know, I know. I am so sorry. I am angry, hurt, lonely. She was my partner. We were supposed to grow old together. Now I am without my best friend. My soul mate. I just hurt." With that Mitch broke down and cried, dropping the beer bottle he was holding and watched it brake on the floor.

"I feel you man. I feel you. But you still have one bright spot. Blair is here man. She needs you so bad. She can't cope with losing her mom and she feels like she lost her dad. You gotta reach out and get her. Now bro. Right now."

"I know. She is hurting too. I will find my daughter." Matt left his brothers and headed home.

Blair found herself looking more and more for the company found in a man's arms. She knew that having those strong arms around her would help give her comfort. She called up James who was more than happy to oblige. He came right over and took her to his place. They talked and then James began to kiss and caress her. Blair felt good being with him and let things flow as they may. That is all the leeway James needed.

His hands began to rub Blair with such tenderness that she had forgotten how much she needed to feel wanted. As she closed her eyes and began to sway James could tell she was losing herself to him. James loved when women were like putty in his hands because he could have his way. He had a reputation of being a womanizer and was very experienced in getting what he wanted from a woman.

Before she knew it Blair was being undressed. First her blouse, then her skirt. James looked at her and began to undress himself.

"Are you sure you ready for this baby?" He asked.

"Yes. Aren't you?"

They spent the night together and Blair fell asleep in his arms. James smiled the whole night feeling he had finally conquered the woman he wanted. Although Blair believed that this man was going to be there for her James had other ideas. He had been a lady's man and saw no need to quit. He was not sensitive to her hurt. So even though Blair had shared her heart with him James still went after other women with Blair unaware of how disloyal her new boyfriend was.

Blair headed home. Her father was not there. So she made herself a sandwich and sat outside looking at the stars. She thought about how different her life would be if she was on a spaceship headed into the galaxy. Mitch came in and saw his daughter staring off into space. He approached her cautiously.

"Hey baby girl. How are you doing?"

"I am okay Papa. How are you?"

"Sorry. So very sorry." With those words Mitch began to tear up.

Blair turned to face him. "Why Papa? What's wrong?"

"I got so caught up in my own pain that I forgot you had pain too. You lost your mom. And your dad couldn't handle it so you lost him too. Well forgive me baby. I am back. Your uncles are here with us as well. And I promise, as long as I have breath in my body, you will never be alone again."

Blair put her head down and began to cry; "Papa, I needed to hear that from you today more than you will ever know. Thank you for loving me."

"Always" With that Matt went and hugged his daughter for a long time. They both cried.
"Let's go get some popcorn. This has been a real emotional movie."

Blair laughed. Then they went into the kitchen and made some popcorn.

She thought that it would be good for her to get away from it all. So joining the GAP became a way out. The Recruiter was glad to get someone like Blair interested in becoming a part of the service. He laid out various career paths she could take and Blair was intrigued. She loved radios and was interested in communication so felt that she would get a lot of practice and develop the skills to be an expert communicator.

Her uncles were all military men although none of them served in GAP. When it became evident that Blair was seriously considering joining the service her uncles were ecstatic. They thought the military would be a welcome distraction for all the pain this family had gone through.

"She is a smart girl. The military will be better with her in it" said Uncle Andy. "I know she can go far and she likes all that electronic stuff. You know they let kids play with radios in space. She will be making calls to Earth from Mars."

"No Andy, they do not. She will learn sub space and space communication aboard both a space station and a space ship. It will be really good for her" said Uncle Stan.

Graduation time was approaching and Blair knew it was time to make a decision. Her first choice was the GAP because she knew she wanted to focus on radio and communications and had heard the support the GA was just what she wanted. She met with her recruiter and finalized her enrollment.

Her father and uncles were all glad that she had enlisted. As always they wanted to offer her words of advice based on their experience in the military.

"Blair, take boot camp seriously. You will need the conditioning to handle being in the field. I know you a woman so you a little weaker than us men but they will expect you to hold your own." said Uncle Andy, Mitch's older brother.

"Uncle, I have no problem keeping up with a man. I am not as strong as a man but I am just as smart and can outrun any man I currently know. I have stamina and I am in great shape."

Stan shared his wisdom with Blair. "The training for communication is excellent. You have a chance to not only learn communication from the bottom up but you will get hands on experience and a chance to work at what you learn. Remember your squad or team will depend on you to connect them with the base whenever you are on a mission."

Sounds great Uncle Stan" said Blair. "I hope to gain a wealth of info so that once I leave the GAP I can get a good job with the skills I have learned."

Blair went to the recruitment meeting. There were about 1000 people there. She was led into a room and given a test unlike any she had ever taken before. It was a simulation where she was an officer and there was a lot of bombing and gunfire going on. Her duty was to fix the radio which had been broken and get a message to her squadron. The situation was interesting.

She got to work immediately. First she tried the radio to see if she could get it to work by making adjustments. When that didn't work she took the tools on the table and opened it up. As soon as she saw it she realized what the problem was and began to repair it. In just a few minutes she had it working again.

She did not know that the whole tests was being viewed by representatives of GAP. They were impressed with how she analyzed the problem with the radio and came up with a game plan to fix it.

"Interesting. She looked at that situation in the same way we would have trained her to. Impressive."

"Yes it is. And she seems like a bright young girl. She should be a fine cadet."

After her meeting Blair met Matt. They listened to the presentation as they sat together. This was there first time being this close to each other. They had no idea it was the start of something new for both of them.

Matt and Blair showed up at their first info meeting at the same time along with a bunch of other recruits. When their eyes met they were both pleasantly surprised by how the other looked. Blair stared at Matt's physique, perfectly formed head and great smile. Matt was taken aback by Blair's long brown hair, beautiful teeth and cute face. These two were like a bunch of deer lost in the headlights. Matt broke the silence.

"Hi, I am Matt."

"Blair. How are you?"

"Great" said Matt. Is this your first time hearing this information? He knew it probably was but he couldn't think of anything else to say.

"Yes it is. How about you?"

"Yeah, me too. Everything looks great. I love this auditorium."

As they walked together they began to look at some of the other students who would be joining them. Not everyone made the GAP squad. Your academics were the beginning along with whether or not you played any sports in high school. All of the students who met the initial requirements to get in were now were being seated in an auditorium for a meeting. There were over 1000 students present.

"Hello, my name is Captain Cornell and I will be your host for this information meeting. I am glad you all signed up for learning more about our GAP program. This initial meeting is to tell you more about who we are, what we are looking for and how you will fit in the overall picture.

Each of you are here because you met our initial qualifications. You have a grade point average above 3.0. You were involved in some sort of sport in high school. You passed your physical exam and showed the mental aptitude to handle our rigorous training for working on a space station. This is Phase I.

Phase 2 will involve going much more in depth. You will be trained physically so that you can endure the rigors of life in space. You will be put on a special diet to boost your immune systems and increase your brain capacity to receive new information. You will be given aptitude tests that will determine where your inherent strengths are so that we can place you in the appropriate department. And you will be tested for our Space Ranger program.

Our Rangers have been taking a beating because we encounter all types of alien life out there. Some are stronger, faster and more intelligent than us and conventional means of law enforcement don't work out here. In order for us to remain competitive we must change how we enforce the law. We have some new developments for the Rangers and this group will be the first to have those enhancements at their disposal.

All in all you picked a great time to join GAP. I want you to read your assignment packets and go where you are instructed. If you have any questions, ask them of the Sergeant who has been assigned to you. That is all."

Matt and Blair read over the info given and were greatly intrigued. "I hope I get picked to be a Ranger. Sounds like why I am here." said Matt.

"Me too. I am so excited to begin."

Neither one of them realized that this chance meeting would be the start of a friendship that would become a love affair. They had a natural chemistry that attracted then to each other but neither one of them was looking for a relationship.

Chapter 4 – Training Camp

The initial 6 weeks of training took place in Arizona USA. It was hot, humid and grueling. The purpose of this first six weeks was to get everyone into shape for boot camp. Unlike some of the other military operations a person could join, if you were picked to take the GAP training you had to be in shape before it began. This means the focus would be on weight training, endurance and getting in the best shape you could. In addition there were changes to the diet implemented since space travel meant eating foods many were not accustomed to & a multitude of brain tests/exercises because being in space could affect the way your brain deals with information.

All the recruits were broken up into squads. Blair and Matt where put on the Alpha Squad. Their training officer was a former Green Beret in incredible shape. He looked at the recruits and laughed. He would talk to other officers and make fun of how his team looked like a team of runaway elephants. He could run a mile in 4 minutes, bench press 300 lbs, do 100 push ups and 500 sit ups with barely breaking a sweat. His handshake was like being grabbed with a pair of vice grips. In addition he held black belts in Karate and Aikido.

Sergeant Simmons got them up each day and for 6 weeks they trained 6 hours a day. The goal was to tone up and drop as much body fat as they could shed in 6 weeks. Each cadet did weight training and was expected to increase their weight lifting by 10% within the 6 week period. Running was instrumental so the group had to get to where they could run 3 miles a day inside of 30 minutes. For some it was a breeze, others had it tough. To keep their minds sharp everyone had 30 minutes of mental exercises each morning and 30 minutes at night.

Sergeant Simmons noticed that the Alpha Team was different than the other teams he had trained. One recruit, Philip Moore, showed uncanny accuracy in shooting rifles.

"How did you get so good with a gun Philip?" Asked Sergeant.

"I grew up on a farm about 10 miles from the nearest town. I didn't have many friends so I got to practice shooting most days. Got so good people began to call me Shooter."

"Oh, is that right? Corporal Smittens, can you join us? And bring 2 rifles with you." The Corporal left the gun range and headed to the Sergeant with 2 rifles.

"Give one to Philip here. Tell me about this gun son."

Philip took the gun in his hands. "Okay. This is the army issued F-116, a single fire highly accurate rifle which fires up to 15 bullets. This one seems light so I would guess it has 3 shells in it."

"Corporal, is he right?"

"Yes sir, even down to 3 shells in the chamber. Loaded it myself. This young man knows his guns."

"Okay Corporal. Give him a target to hit from here."

"Sir, we are not at the range. It's against regulations to discharge a gun outside of the range."

"Like I said. Give him a target. How far is this gun accurate to without a scope?'

"About 100 yards."

"Okay Shooter, do you see that Red and White Bullseye at the archery range over there. No one is using it now. I want you to try to hit the bulls eye."

"OK"

Everyone started walking to the archery range. Philip didn't move.

'Come on. Let's go."

" I thought you wanted me to hit it from here."

"Son, that is probably more than 100m yards away. We can get closer."

Philip held the gun up to his eyes." I can see the bulls eye from here. It's a little over 100 yards but I think we can still hit the target."

The Corporal looked at Philip. "You honestly think you can hit the target from here?'

"Yes sir, I think I can."

"Do it" said the Sergeant.

Philip stepped back, scoped out the target, handled the gun to gauge its weight, looked to the wind and then took aim. He fired.

"From this distance we cannot tell if you hit it or not. What do you think?

"I did. Would you like me to do it again. Or use all the bullets in this gun?"

"Yes."

Philip took aim and fired twice more. Then they all began to walk towards the bulls eye. Philip hit the target all 3 times in a straight line.
'Freaking incredible. Just incredible"said the Corporal.

"That is all Corporal."

"You have a gift son. And we are going to help you to use it."

The next morning was hand to hand combat. While students were giving instructions Sergeant noticed that one cadet, Danny Simpson, was not very interested. He walked up to him.

"What, this is too slow for you son?'

"No, it's just that I have been studying martial arts since I was a kid. This is very rudimentary to me."

'Oh it is, is it. So let's spice it up a bit. Young, Fuller and Mitchell. On the mat. Danny, you have 3 of our training students here, each proficient in his own right in a particular martial arts discipline. Can you spar with them?"

"I can sir but I don't want to hurt anyone."

"I think they can handle it. What do you say boys?" They all agreed.

Young here focuses on Karate. He will be first. He attacked Danny with some moves which Danny gracefully avoided. Then Danny blocked and stopped him by putting him to the floor.

'Good Danny. Here is Fuller, his discipline is Aikido. Take him." With that Fuller attacked. Danny fought him using Aikido and used a wrist lock to flip him over.

"Mitchell studied kung fu. Attack." Mitchel assumed the Preying Mantis stance. Danny countered with the snake maneuver. He kicked Mitchell and he was down.

"Very good. Now, all 3 attack at once. I want to see what style he uses when there are multiple attackers." They all surrounded Danny. He adopted the Crane Style and took out Young. Then he switched to hard karate and took out Fuller. When he faced Mitchell he switched to Aikido and took him out.

"That young man, was most impressive. How in the world did you learn so many styles of fighting?"

"My father trained me since I was 6 years old. He required a different black belt every 3 years. I currently have belts in Karate, Aikido, Kung fu and Jiu Jitsu.

"You are a walking Ninja."

"Philip spent his time shooting guns. I spent mine learning how to fight."

At the end of the 6 weeks they had all lost weight and their muscle definition was showing. Now they were ready to board a shuttle and head to the space station.

The trip to the station was an exciting one. No one on the transport had ever been in space so this was a real out of earth adventure. The ride took about 6 hours to make it to the station. The sights to behold outside of the ship were breath taking.

Matt and Blair were becoming closer and closer. The found they had many things in common. Matt said "look Blair, this is a pretty serious position we are applying for. Are you sure you want to roam the universe chasing bad guys?"

Blair said "I know. It is scary. But the chance for scientific exploration is awesome. I want to be a part of what is going on in space."

As they approached the station they say a battle cruiser in orbit around the station. Blair asked "what ship is that/"

'That, my eager cadets, is ALMA. She is the Space Rangers flagship, just off the presses. She will be ready for flight in a few weeks and the Rangers will have an AI ship to assist them in their duties. ALMA is totally automated and can function manually or on auto pilot. She also has a voice simulator so you can talk to her."

"Awesome."

They got off the shuttle and were checked into their rooms. Training Phase II would start at 4 am. Getting up at 4:00 am each morning would some getting used to. Their training officer, Sergeant Simmons was a rough character. He pushed his recruits harder than anyone else and expected them to perform at a high level.

"Good morning team. Let's get off on a good foot. I am Sergeant Simmons and my job is to get you all combat ready. I know some of you have come here and played sports in school. Or you may be one of those who went to the gym and worked out often. Whatever the case my job is to break you to make you. So I am going to push you beyond any limits you have ever had before and I will cause you to see yourself doing things you never thought possible. I have had team members leave her doing 100 push ups, 1000 sit-ups and running a mile inside of 5 minutes. Some could hold their breath underwater for 1 minute with no problem. These men and women left me as soldiers. Do you want to leave me as a soldier"?

"Yes we do" the team responded.

"Yes we do sir. Every time you talk to me, end your statement with the word sir. Understand"?

"Yes sir".

"What did you say? I can't hear you"?

"YES SIR".

And so it began. Running 3 miles each day, doing as many push-ups and sit-ups as we could. There were cardio exercises, stretching exercises and running all the time. This was clearly tougher than any sports event the team had been a part of before. And we felt ourselves buckling under the pressure.

Lunch time. Plenty of food that was good for the body. No carbonated beverages. Each team member was expected to eat a balanced meal when told. The team did everything together from 4 am till 4pm each day. The early evening was to relax and get some rest. At 8 pm the team assembled before the Sergeant and practiced whatever exercise he had designed for that day.

We were given shots daily. They were supposed to boost our immune system so that we could travel in space. What we did not know was that they did other things too. This was to be revealed to us at a later time. In addition they were reading required to get ready for the many expectations new soldiers were to deliver. The Sergeant was a no nonsense officer and expected no less than their absolute best from his team members.

During our breaks we had a few visitors.

"Hello recruits. My name is Dr Sagen. I will be your classroom instructor. My job is to prepare you mentally to become a part of our station. There will be required assignments due every Friday no exception. You will have one week to study. It is advised you study in a group setting as this will help you to retain what you are learning. Your assignments are downloaded to your pads. I would suggest you begin today."

"Hello. My name is Dr Mellows. I will be the one examining you and running tests to see how you are doing physically. I need you to make sure you are eating what you are given, taking all your vitamins and you will be getting some shots on occasion. Try to cooperate with me, okay?"

As the week progressed it became clear that none of the new recruits were in the physical shape expected of them. They soon began to grumble and complain causing the Sergeant to go off on them.

"I know we have girls and boys on this team but with all this whining I can't tell which is which. Seems like you all need dresses and jump ropes. I don't know what you expected when you signed up for the Army but here we make a difference. So quit your gripping, man up and work out. In time your body will adjust. Your mind needs to make the adjustments now."

The first week of boot camp was officially over. In the barracks the team complained about everything. They soon got a reputation for being whiners. But not Blair. She stuck it out and gave it 110%.

"How can you be so positive with all this stuff the Sergeant puts us through" asked Willie, a fellow private. "Because I lost my mom a year ago and that was clearly the hardest thing anyone could ever go through. I won't let something like boot camp beat me when cancer couldn't".

Willie was speechless. "I am so sorry Blair. I did not know. I will stop complaining so much as well."

One by one the team agreed to stop complaining. Then Matthew said "Blair, we never know what problems people have gone through before we met them. I am glad to see how you can be so positive with the negative things you had to face. You are an inspiration to us all, me first."

Blair smiled. "Thank you."

The senior brass held a meeting that Saturday following the first week to go over results. These meetings were chaired by the Admiral, commander of the Space Ranger Program.

"Give me your updates on our new recruits"

"Well, at boot camp they have been a bunch of little whiners, which is expected. From what I can gather the woman Blair brought Team Alpha back in line with her perspective as to how they should be handling things. I found Matt offered support and together they rallied the Team back to focusing on what is important."

"Great Sergeant. Keep me posted."

"I have performed the academic pretests. So far it looks like we are still missing a few key components to completing the assignment you have for this Space Ranger team. I have not identified who will be our engineer, pilot and weapons expert. But it is still early."

"Dr Sagen I need to have some candidates by next week. We have to begin their training."

"Okay you shall have it."

"I have given the recruits physicals and you have my results. So far all is looking good but I have not found the particular biochemistry match we need yet for the Ranger program. However I believe that 6 of the candidates in Alpha Squad may be suitable once we see their results to treatments."

"Dr Mellows, you know that is key to making this all work. I trust you will have those results within a week?"

"I will push them through for your review."

"Great. Dr Siloquin, explain to us all again how the Space Ranger program enhancements will work."

"Gladly." said Dr. Siloquin. He was the bio chemist who developed the "implant."

"Once we are sure we have the right specimen we will surgically implant them with an enhancement module. This module will activate certain genetic instructions within the candidate. The goal is to increase the physical and mental state of the specimen. This will result in increased strength, muscle coordination, reflexes, and brain power. We will activate the module through the use of an implant into the specimens fingers and the Space Ranger badge they are assigned. Now we do run the risk of having candidates who already have certain inherent abilities which may be dormant in them now. Only tests will tell."

"Such as what?" asked the Admiral?

"Well, they have telepathic abilities. Or untapped reservoirs of strength. We won't know until we can get the results back."

"Great Team. Keep me posted. We meet back in one week. I want to have my team of 6 by month's end and surgery to follow the next week."

In a week the Alpha Team was all cleared. They would be one of 6 new Space Ranger teams to hit the galaxy. The process to get ready began.

Later that night Matt was having a night cap in the Station Lounge called "Star Stopper." While sipping on a coke Blair walked into the room. She saw him and came and sat at his table.

"Well, well, well, If it is not our future Captain. How are you tonight sir?" said Blair.

"I am well. And I must say that you are looking rather stunning tonight as well."

"Oh, so you notice when a lady dresses up? I thought you were working out so much that your eyesight was clouded. I mean, what's a girl to do to be noticed on a station with a bunch of men."

"I did not know you were looking to get noticed. Any reason why?"

"I would like to have some company when I come to the Star Stopper. We all need ways to unwind."

"Well you do know that fraternizing with soldiers is forbidden. I mean, I think you would be fun to fraternize with and all."

"So you want to get with me huh? Captain, whatever for?"

"I think I had better watch what I say. You seem like you will hold me to my words."

"I certainly would. I don't like when men lie to me."

"Seems like the lady has been lied to before. There has to be an interesting story to that one."

"Let's just say I gave my heart and soul to a man before joining GAP and he proved to me why a woman is foolish to trust a man. Men only want one thing from you and once they get it, you are done."

"Seriously, I would like to hear the details. All men are not dogs out on the hunt. Some of us are actually pretty nice guys."

Blair smiled at Matt. She realized she was being a bit hard and he had done nothing to deserve it. So she softened up a bit.

"I went through a hard time with my mom passing before I came to GAP. My boyfriend at the time was the one person I felt I could turn to for comfort. He wanted comfort all right. But what he didn't want was to be bothered with my problems. Oh well, who needs men anyway?"

"I left a really good girl when I graduated high school to come here. She understood my decision but it was very hard. We broke up so that she would not have to feel guilty waiting on me. "

"At least you all went through the emotional break up. That is the way it's supposed to be. You cared for each other and it hurts when you have to leave. My nut of a boyfriend was only interested in getting with as many girls as he could."

"I don't know the guy but I can say that any man who has the chance to be with you should not be so stupid as to throw it all away. A good woman does not come along all the time."

"So now I am a good woman? Boy, you can lay the lines on can't you?"

"No lines. But since we are talking about what we are doing, you could pull those daggers out of your hands. Not every man is interested in hurting you. Some of us are even pretty nice."

"Umm huh. Okay, we shall see."

"Blair, I know we are going to be working together and I know we all have a past we are bringing here. I want to get to know you because we are going to be spending a lot of time. And I don't know; I think you are an awesome person who is putting on this armor to protect her heart. If I am right I hope the time will come when you take off the armor and let a fella get back in that heart of yours again."

Blair smiled and said nothing. The words stung a bit but she knew he was right. She had never really dealt with her hurt from the break up. She tried to ignore it and believe it would be okay. Yet she did care for him and she needed to process a change. Little did she know that her time would be coming really soon.

There were a lot of tests coming up for the young cadets. Today they were meeting with Dr. Sagen who was the one administering the psychological exams. He spoke to every cadet in the program at some time or another. He knew the GAP program was looking for certain types of people to become Space Rangers. He called them into his office one at a time.

"Matt, good to finally get to sit down and talk with you. We are going to review your profile and have a discussion. You ready?"

"Yes sir, I am."

"Fine. So I see you came for a very stable family with both parents at home and siblings. You were a great student and you even participated in athletics."

"Yes sir."

"So what would you say is the biggest obstacle that Matthew needs to overcome in his life right now?"

Matt sat up and thought for a minute. Then he spoke. "I guess when I set my mind to do or get something I expect instant results but I don't spend enough time viewing all the options available to me that can give me the same result."

"You feel impatient."

"Yes, I am.

"Okay, what do you plan on doing with that knowledge?'

"Don't know yet Dr. Sagen. Working on it."

"That is good. The only way to lean patience is to be put in situations where you have no choice but to exercise patience. The patience muscle grows as it gets exercised."

"But it has been my experience that people in those situation cannot change their outcome. That would drive me crazy."

"So let me ask you: are you looking to control your outcomes or learn how to wait until the outcomes appear?"

"I hate to say control them but I think I am. I don't see myself as a controlling person but I do know that I hate when my destiny is in someone else hands. I like to control how I live my life and where it is going."

"I see."

"I realize there are some things out of our control and we just have to deal with them. I struggle big time with those things but have to learn how to accept them. I think I am more of a possibility thinker. I see solutions to all problems. You just have to find them."

'Have you always been this way?"

"Yes I have. My parents use to tell me that whenever someone said the word "can't" or "impossible" around me I would light up. I cannot imagine being in a place where no alternatives exist."

"Being in GAP may be a challenge for you. Granted change is a way of life around here. But the powers that be are in power and you cannot change that just because you happen to not like it."

"I realize that is the culture in the military and you have go along with it."

"Yes it is. So tell me, why do you envision yourself leading a team?"

"I have always been a person of influence. And I am a quick decision maker. I think these two qualities will fit me nicely commanding my own team. "

"And what is your future goal?"

"I want to one day command a squadron and have a whole fleet of ships under my command."

"That is a big dream indeed. Do you think you can pull that off?"

"Yes sir, I think I can."

'Well thank you Matt. That is all for now." With that Matt departed the doctor's office. He found the questions to be a little strange but just shook his head and went on.

Dr. Sagen motioned for Blair to follow him next.

"Blair, good to get this time with you. We will be reviewing your profile, your past and what your future goals are. You game?"

'Sounds great."

"I see that you were an only child. You did great in school and participated in athletics. You even went to State a few times. Very impressive."

"Thank you."

"I also see that your mother passed away while you were still in high school. I am sorry. It must have been very hard and your father."

"Extremely."

"Want to talk about it?"

"Well, we can. I am not over it but I have accepted it."

"You will never be truly over it. What you will do is grow each day to be able to deal with it and manage the hurt reliving the feelings will cause."

"I see."

"What happened to you afterwards?"

"What makes you think something happened?"

"You cannot go through such a traumatic experience and not have some repercussions. Not saying anything bad happened. But something did."

"Oh Doc, you got about 2 days?"

"That much?"

"Yeah. I kind of went on an emotional roller coaster ride. My dad could not handle losing mom so he drank and avoided me and his brothers. I had no one to talk to so I turned to my boyfriend for comfort but he was a detached jerk so I ended up feeling even worse after being with him. Fortunately my father did come around and we cried and held onto each other until we both could cope with it."

"I see."

"I guess everyone has a story to tell huh?"

"Yes they do. So what did you learn from that experience that you feel will make you a better person?'

"I learned that life is short and we must live each day. Tomorrow isn't promised so you cannot be so far into your future that you forget you have to live today. I will cherish the people in my life who I come to love. I will treat them as special because they are."

"I see. Is there anyone special in your life right now?"

Blair paused. 'No, not really at this time."

"Relationships can be really hard in the military. And you will be a woman on a team with mostly men so I caution you to guard your heart. Love has a strange way of messing things up when you are on a mission."

"No worries doc. I don't plan on falling in love anytime soon."

"And what are your future plans?"

"I want to one day teach communications at GAP academy. I think it is vital to a mission's success."

"I see. Thank you Blair. Our time is up. I will be speaking to you again sometime in the future."

"Thank you doc." With that Blair left Dr Sagan's office.

Dr Sagen went into the reception room and invited Philip to join him.

"Philip, I have your profile and we are going to go over it. I need to discuss your results, your past and what your future plans are. You ready?"

"Sure am."

"I see here you grew up on a farm. Did you have siblings?"

"Yes, 3 sisters and 2 brothers."

"Busy bunch there. Guess you all grew up pretty close."

"One of brothers was 10 years older than me. He joined the military and was killed in action."

'Oh, sorry to hear that."

"Thank you. He died saving his squad. My brother was a heck of a shot. When the enemy drones attacked he took his rifle and held them off until his men could get to safety. One of the drones detonated near him and that's how he died."

"I see. How does it make you feel thinking about your brother?"

"Well I know he saved their lives because he could shoot. I said I want to be like Ricky when I grow up so I practiced and practiced till I could hit shots he use to hit. My brother was still better than me but I am coming up close."

"Congratulations. You took a negative experience and used it to propel yourself to a positive place in this life. Good job."

"Thank you."

"What is your future goals? What do you want to do with your military career?"

"I want to move up the ranks and share my shooting knowledge with cadets to better prepare them for being in the field."

"Oh you want to teach shooting in the academy?"

"Yes, if they will have me."

"You keep on doing what you are doing. You are on the right road and your destination is clear. You should hit it."

'Thanks sir."

"Philip, that is it. Thank you for coming in." With that Philip departed.

Dr Sagen motioned for Danny to join him. Now Danny was a man of few words. He spoke when spoken to but rarely offered conversation unless someone else initiated it.

"Danny, good to see you. Make yourself comfortable. We are going to be discussing some items in your profile and then asking questions about your past and how you ended up here. Everything we discuss is confidential as far as content goes. Are you ready?"

"Yes."

"I strive to give 110% in whatever endeavor I am involved in. That is reflected in how I have performed thus far."

"So you are a very disciplined person by nature?"

"Yes."

"Is that how you were raised? With much structure and order?"

"Yes."

"Please explain further for me."

"My parents were both very logical people who focused on controlling their minds and bodies. My father was a martial artist who competed on a world level. He is my example of how to coordinate spirit, soul and body."

'Your father trained you to be like this?"

'Yes."

"And I understand that you hold belts in Karate, Aikido, Kung Fu and Jui Jitsu. Is that correct?"

"Yes it is."

"And which belt do you hold in each of those"?'

"Black belt."

"Why did you get a black belt in 4 martial arts disciplines? That must have been incredibly tough."

"It was. But I am glad to have done that because it gave me discipline which helps me greatly today."

"I see. So what do you see as some of the weaknesses you possess?'

"I focus on my strengths but I am aware of my weaknesses. I am not the most patient person, I do enjoy having fun and I like watching movies."

"Okay, well that's okay. Why movies?"

"My father did not believe in taking time to do anything but practice. So when I left home I found that movies relax me. I have incorporated watching them into my daily activities."

"Any types of movies in particular?"

'I enjoy drama, comedy and action. I stay away from horror."

"So you feel that you need to develop patience too?'

'No I am a patient person. But when you reach the height of enlightenment you are patient in all things. I strive to arrive there but have not arrived there yet.

"I see. Anything else?'

"Yes. I perceive that you are looking for something from me but as of yet I cannot ascertain what that is. Please come straight out and ask me."

Dr Sagen was stunned by Danny's perception of their meeting. He was careful to not reveal any clues and still Danny caught on quickly.

"Okay. I am wondering why someone with your background would want to join the military. "

"I abhor violence and I hate when those who have power abuse it to enslave others. The DASHOOOR are ruthless villains who will stop at nothing unless they are stopped. It will take people like me to make that happen. "

"That makes sense. What do you mean, people like you?"

"Those who want to see an evil tyrant's rule end."

"And where do you see your future taking you?"

"I wish to teach self defense at the academy."

"Great goal. We could sure use someone like you."

"Thank you doctor. Your words are kind."

"That will conclude our business for today. I will get back with you at a later time." With that Danny departed.

"Lin, come on in and take a seat. We are going to discuss your profile, your past and what your place would be in the GAP program. Are you ready?"

"Yes, I am ready Dr Sagen."

"You graduated in the top 2% of your class. You have degrees in medicine and biology. What were your career plans?"

"I wanted to engage in research. Space travel was not at the top of my list."

"So why are you here?"

"I was recruited and told that space exploration was the best place to do research. I would be given state of the art equipment and a whole scientific team to work with. It was too much to turn down."

"You are excited about science. Well this is a great place to share that excitement. We have learned much about the galaxy since we joined GAP. And there is so much yet to be explored."

"That is why I am here Dr Sagen."

"I see from your profile that both of your parents are medical doctors. They must be proud to see you moving in their chosen profession."

"Yeah, you could say that."

"I note a bit of sarcasm with your answer. Care to elaborate."

"Let's just say that my parents love when you follow the direction they set in the way they set it. I have traveled down another road and yet ended up here nonetheless. I wanted to be a researcher and my parents were livid to say the least. They felt I was wasting my time with something any moron could do. They failed to see that research was my passion because it sparked my curiosity of what could be. Eventually I turned to medical research and although they were glad I focused on medicine they were still upset with the research concept."

"I see. How did you respond to all of this from your parents?'

"I have learned that the best I could hope to do was to do my best. Ultimately it is my life and I have to live it the way I choose. They hate that concept but for the sake of a relationship with me they accept it."

'So you learned early on how to persevere in spite of adversity. An admiral trait for someone engaged in space travel. Trust me you will have plenty of adversity there."

"I think so. But I feel I am prepared and the man to do the job."

"I think you are right. When you get a chance to see the research lab assembled for the Space Rangers I think you will be well pleased."

"I am looking forward to it."

"What are your future plans?"

"One day I want to work at the station in the research department. I hope to make a contribution here."

"That concludes our meeting for today Lin. I will be in touch." With that Lin departed.

Art was the last one to come in and meet Dr. Sagen. He was the only ranger candidate that Dr. Sagen had some issues with. Based on his test results Art showed a high aptitude for mental intuitiveness. The challenge was there was nothing in his background to suggest where this came from. As Dr. Sagen reviewed his family history he discovered that Art's mother had a shaky past. It was almost impossible to determine where she came from. He met Art in the lobby.

'Art, you are my last candidate for today. Your other team members survived so it won't be too hard for you. I will be going over your profile and asking questions about your past. We also will look at why you chose to be a part of GAP. Okay?"

"Yes, that is fine."

"I see here that you were studying Astronomy in school. What were your plans?"

"To work with NASA and deal with space. I love studying stars and planets."

"Wow. Is that why you are here today?"

"Yes."

"When I look at your past I see that you were adopted. Do you know anything about your birth parents?"

"No, I never met them."

"Are they still alive?"

" I do not know. I have never tried to find out."

"Why?"

"Figured, why should I?"

"Many people in your situation look for their parents to gain a connection. Did you ever want to do that?"

"No."

"Why?"

"I was raised by the Wolfe family. They were loving, accepting and showed me that I had worth and value. They parented me so that I missed nothing growing up."

"That is good. As you know we do background checks on all GAP candidates. The purpose is to find out as much about where you come from as possible to determine how you will deal with the dynamics of prolonged space travel. I must say we have a few questions about your past."

'What questions Doc?"

"For starters, your blood work shows some unusual anomalies You have an incredibly intuitive mind. Did you notice you were different from other children as you grew up?"

"Yes I was teased and called weird."

"Why?"

"I don't know. I would be playing with them and try to fit in. But someone I would know or sense when one of them was trying to do something they should not be doing. This always caused problems and eventually the kids would kick me out of the group. I guess that's one reason I turned to the stars. They never rejected me."

"We will be conducting some more tests. The reason for this is as we start you on the GAP treatments you will undoubtedly respond differently than your team mates. I want that to be a positive experience. "

"Thank you Doctor."

"Anything else?"

"No, I have no further questions."

"See you later then." With that Art left.

Chapter 5 – The Encounters

As the weeks progressed the teams became better and better. They were given small assignments to build teamwork. Matthew rose up quickly through the ranks. He was a smart leader and found that people followed him naturally. As he began to accept more and more responsibility the officers noticed his leadership. They began to recommend him for certain assignments. One officer, Captain Contagion, took Matthew under his wing and began to show him how to lead those recruits who would soon be put under his care.

They would spend hours just talking about how to move people to do what was necessary to get the job done. Matthew became a sponge who soaked up all the info he could get out of the Captain. He wanted to be the best soldier he could be. He knew that as long as he had a mentor he was on the right path.

"Protect your men. Serve them by keeping their best interests at heart. But never forego the mission. It is the mission that binds you all together. You are united for a purpose; to get a job done. Help your men to remember that."

"Yes sir, I will" said Matthew. He wanted to motivate his team to do their very best and support each other in all endeavors. Unity, brotherhood, togetherness, all rolled up into one.

Later that night as Matt was headed back to his barracks he saw Blair taking an evening run. He stopped her.

"Hey you. What are you doing? Don't we get enough exercise already?"

"Yes we do but I am use to running 5 miles a day so I try to maintain that regiment. What are you doing out here this late? asked Blair.

"Weekly meeting with the Captain. Just getting some mentoring done."

"Oh lucky you. I am sure all this training will make you great."

As they both stood there staring at each other, it became an awkward moment. Matt was thinking to himself about how nice Blair looked while she was staring at him. Matt broke the silence.

"Ah, I think I had better turn in. Good night." With that Matt turned and walked away. Blair ran home and thought about how Matt did look good to her. She was beginning to fall for him.

Early the next morning Matt got up to spend some time meditating and running. He liked the quietness of the day to get his head around what he needed to do. As soon as he understood that he got more done after finding quiet time than by bypassing it he opted to find that quiet place each morning. He was contemplating his future as a GAP soldier. He was also thinking about Blair.

Something about her really caught his attention. Maybe it was the way she wore her hair. He loved the way it made the rest of her facial features look. Her eyes sparkled in the light and her smile was magic. To top it off he thought she was had a body that caught everyone's eye when she walked by. He knew he was there to focus on the program but he had a hard time when all he could do is think about her.

He wondered if she was thinking about him. He hoped she was.

After spending 30 minutes in reflection Matt got up and did a run for a mile to get ready for the day.

Blair was tired. She had not slept much because she was dreaming about Matt. She went to sleep imagining his cologne. The smell was driving her wild. She thought about how cute his face was, how broad his shoulders were under his shirt and the fact that he smiled a lot whenever he talked. She liked his smile.

How could she do this? How could she focus on a man when there was so much more to look into? She knew that her training would require single focus. Matt was becoming a distraction. Logic told her to ignore him. Her heart said something different.

Chapter 6 - Becoming a Space Ranger

The GAP General Assembly met to discuss the affairs going on in their regions. It was made up of 8 interplanetary groups from 8 different galaxies. Each group lead a team of Rangers that was responsible for their galaxy. The Earth Alliance was represented by Admiral Thomas Pratt. He voted on policy and helped the General Assembly to enforce intergalactic laws.

Gimorralli, the commander for the Xlanin planet, led the meeting; "Let this meeting come to order. Now we all have looked at the report and the crimes being committed in our regions. We have reason to believe it is the crime syndicate "DAASHOR" that is trying to establish a foothold in the galaxies. Everyone is looking to the Rangers to put a stop to this. I am asking each of you to ramp up your efforts and get these criminal locked up."

"I agree. When we meet again we can give an account of the progress we are making." said Sammuol, General for the planet Nephilis.

On Earth there was an enhanced Space Ranger program which meant there needed to be a whole new group. The General got his list of recruits. Each came with a set of abilities that made them unique. They also possessed the gene structure necessary for the implant to work as it was designed. This part of the program has not been shared with anyone outside of the department that developed the software and hardware that was to be used.

The Admiral had his six recruits brought to him along with his Space Rangers support team. This meeting was to be the one that would determine the direction the program would go in.

"Matthew Lincoln, Blair Hutchinson, Philip Moore, Danny Simpson, Lin Su and Art Wolfe. You six have been chosen because you have passed the initial boot camp, all of our preliminary tests and our psychological evaluations. Now I need to tell you that from this point on you are going to be privy to top secret information that is not shared with anyone outside of this room. Before I get into the details I have to tell you two things; first, you can refuse to be a part of this initiative and return to GAP. You will be assigned a department and your duties will be given to you. There is no problem refusing this assignment. Second, if you stay, you have to agree to every component of the program, no questions asked. Some of your questions will not be answered and you will have to be okay with that. If you say yes, there is no turning back, not changes. Your life will never be the same and neither will you. I will give you a few minutes to decide. Dr Sagen, show them into the conference room." said the Admiral.

"This way recruits." said the Admiral's aide.

Once they were in the room each recruit was given a disclosure to read. It gave them a chance to back out of the program.

"Well guys, let's talk about this. I will respect any decision any of you make. For me I came here to be a part of GAP and wanted to enroll in the Space Ranger program. I know it is a top secret project but I am interested in going all the way."

"Me, too. I joined to work for GAP and I am ready to go the distance." said Philip.

"I am unsure" said Art. "I would like more details before committing."

"Well that's not going to happen. Apparently you have to keep this stuff secret so you have to commit." said Blair." I agree with Matt. I am all in."

"So am I. I think that if we expect to be a part of an interstellar police force, we are going to need all the help we can get." said Danny.

All the while Lin Su sat in silence. He was afraid of letting them know the truth about why he was here.

"Okay Art and Lin. You two need to decide. Otherwise us four will give our commitment and see who else wants to join the team."

"I am in" said Art.

"So am I" said Lin Su.

The Admiral brought everyone together and saw they all committed to the program. He smiled.

"Let's get this show on the road. We will now explain to you in detail how the Space Ranger program works. Dr Siloquin, as the leader of this research team give our recruits the information" said the Admiral.

"The GAP program was started 10 years ago with the expansion of space exploration. Earth joined and we have struggled with the conditions we found in deep space. Our bodies are not equipped to handle deep space travel. So if we are going to be an effective member of GAP we need some enhancement.

We have developed an implant that will enhance certain human abilities. This implant must be surgically inserted into the subject. Once inserted it will activate by a combination of two components. Once is a chip that will be implanted into the finger. The other is a badge each Ranger will wear. When the chip and badge are bought together the subject concentrates and the implant will activate.

In addition to these human enhancements we have been treating all of you soldiers with our enhancement serum. By now your bodies have adjusted to the serum and once the implant is inserted the transformation is complete.

You will not look or feel different. But you will have enhanced strength, speed, reflexes and your mind will be sharper. The effects will last for a short period of time but you can reactivate as you need to.

Now the tricky part of this is that each of you has certain biological and physical traits that are unique to you. I would like to speculate how the implant will work in you but until it is surgically implanted and tested, we won't know."

"So we will wake up as freaks with something in our brains?" said Art.

"You will wake up and not be freaks. But you will never be the way you are right now again."

"Okay, so what's the next step." Said Phillip?

"Dr Merrill will be conducting the surgery on each of you. Once it is completed we will need at least 3 days before the implant will have joined with the chemical adjustments in your body."

"Let's get this show on the road. I am curious as to what will happen to me." said Art.

"All right. We will schedule the surgeries tomorrow. Dr Merrill, get things ready."

"We will begin at 7 am tomorrow. Should only take 2 hours per subject."

The surgeries went fine. The progress of the group was monitored closely. After the 3 day window had passed a series of tests were conducted.

The Admiral brought his team together for another meeting. "Give me a progress report."

Dr Merrill said "The surgeries went fine for all 6 candidates. The implants have been successfully inserted and their bodies are suffering no ill effects after 72 hours. It's a go.

Dr Siloquin said "I have closely monitored their blood work and biochemical responses to have the implant interact with the serum enhancements they have been taking. We moved to using an IV for the last 3 days to give them the last treatment. I have looked at their blood work and the adjustments have begun."

Dr. Mellows said "We have had them on the special diet for over 6 months and their bodies have made all adjustments permanent. They are ready to go."

Dr. Sagen said "The psychological profiles have all been completed. Based on what I see we may have a problem. I believe one of the recruits is exhibiting changes in his thinking. There is some type of transformation going on with him. I fear he will not come away the same as when we started."

"Which recruit are you referring to?" asked the General.

"Art Wolfe. His mental state seems to be in a state of turmoil. Hard to understand why when I look at his medical records. He should have adjusted easily."

"What do you suggest?"

"I need to meet with him and do a more in-depth psychological analysis. "

The team was all bought together and told what would happen next.

"It looks like the implants were successfully inserted. We need to run tests with you all geared up. I would like to explain to you the enhancements we have for each of you" says Dr Than.

Dr Than designed the badges, implants and worked with the team that developed the weaponry for the Space Rangers. "Now remember Rangers, each implant will enhance your normal strength, speed and reflexes. There are some other enhancements that impact each of you individually. "

He began with Matthew.

You are being given 3 weapons to help you in your law enforcement.

-
 -
  -

ALMA – Space Ranger ship which is the highest form of A.I. in existence. The ship communicates with the team and is able to reason in decision making. The ship will not override the crew's directives and follows the Captain above all. Space Avenger pilots the ship and can communicate with Alma telepathically. The ship has reflector shields, laser guns, torpedoes and rapid fire explosive pellets. The ship carries a year's supply of goods to allow it to spend months in space without needing to refuel or get supplies.

One day the Captain contacted Matthew to give him an update of what was going on in a region Matt had worked in before. The info about what was going on in the region was of utmost interest to our national security.

They sat down to discuss things over a cup of coffee.

"Matt what I am going to tell you is off the record. Once the mission is assigned to a team they will be briefed. I felt it appropriate to speak to you in this way because I think this is something right up the alley of your team."

"What is it sir?" asked Matt.

"We believe there is a group of rebels operating in the region. This group has been growing in number and when we sent in a group of soldiers to survey the situation they were captured. We have to mount a rescue attempt within the next 48 or we run the risk of them being moved. "It gets more complicated. One of the soldiers being held is the son of Sergeant Simmons, your old trainer. He wants to go in but I forbade it. I told him I had another unit in mind."

"I see sir."

"There is a planet on the far side of the region called Plastar. We believe a top illegal weapons dealer is hiding there. This man is powerful and has a small army at his command. He sells weapons to the rebels. What we need to do is capture him and discover what the plans of his buyer are. Your objective will be to capture him and return him here for questioning. We believe this is a front for the new intergalactic crime syndicate."

"Who is the mastermind behind this sir?"

"We are not sure as to how far this all goes up the ladder. What we do know is we need to get the General into our possession and question him. I need you to go get him."

"Okay sir. Tell me about his current location."

"He is on a planet called Plastar, about 4 days outside of the Milky Way. The planet has compatible atmosphere to Earth. He has a regiment of armed guards. I believe your team can go in and take him. We need to bring him to us within the next 10 days. So you will find the 2 American prisoners and bring the General back to us."

"When do we leave?"

"in 2 days. Get your team ready."

"Yes sir."

Matt assembled his team. "We have our mission. The ship leaves in 2 days. Art I need you to make sure we are ready for intergalactic flight. Lin Su, medical lab is yours. Philip and Danny, get our weapons loaded on board. Check the Shuttle to make sure it is ready as well. Blair, set up Alma for ship to station communication. Alma, I know you can hear me. Get me information on our meeting along with details about the planet we are visiting."

We took off in 2 days and was headed to Plastar. In Hyperspace drive we could get there inside of 4 days.

Matt called a meeting. Alma had the ship on auto pilot.

"Okay team here is the planet we are going to. From here on out as we are on missions, we use our code names. Shooter, you will explore our options and keep us under surveillance from a distance. Space, you and Alma will monitor the planet and keep us posted. I want to know the minute anything changes. Blair and Doc, you are the information team. I want to know what they know about us and their future plans. Ninja and I will work the extraction with support from Philip. If things go haywire inside, Blair and Doc become our back up. Any questions?"

"No"

"Then suit up."

It was a cold, gloomy day on Plastar. This planet was notorious for fog, cold weather and unclear skies. We tried to not pay attention to how uncomfortable we all were but it was pretty difficult. Nevertheless we had a mission and that came first.

We had just gotten a lead from Central Command that Bishkek was here. He was a smuggler know for getting whatever you wanted in this part of the galaxy. We figure that he would be the person the General would contact. He was known to be resourceful and well connected to the underworld in space.

Once our ship entered into orbit we left Space on the ship. The team had their assignment. Philip took a jet pack to travel around. The rest of us took the Shuttle.

There was a town about 2 miles from the compound. To survey the situation and to see if anyone saw Bishkek, Captain Combat and Lady Hawk went into town, dressed in a long overcoat. Ninja tagged along but hid in the shadows to be a support if things went south.

Once inside, all the men and women looked at the two strangers in their midst.

"A goofy looking Martian approached them and said "is the woman for sale? I hear earth women are freaky."

"Yes she is but you can't afford her. We only speak to Bishkek I hear he gives a good price for merchandise."

"Bishkek will not speak to you. You are not a trader. He does not deal with strangers."

"Even if I can offer him this?" Matt took his Multi gun out and showed to the Martian.

'He has enough guns.'

"But not like this one. It can do some great things. Wanna see?"

'Sure"

"But I will only demonstrate for Bishkek If I show you and he is not here, I will be forced to kill you." With that Matt pointed his Multi gun and Bishkek and said "357 Magnum". The gun made a noise and the Martian was afraid.

"What does that mean?"

"If you fail to bring in Bishkek, this gun will fill you with the bullets from a 357 Magnum. It can leave a hole in you the size of that plate on the counter."

"You are bluffing."

"Maybe, but if not, only you will die here. No one else. You ready to risk your life?"

The Martian smiled, and then his smile turned into a frown. Now he was afraid.

"Don't kill me man. I only wanted your woman."

"Bishkek Now."

The martian left to make a call.

"Bishkek will see you at 6 tonight."

"We will be here at 6."

Chapter 7-The Capture

Alma took the Space Rangers to the compound and landed on the airstrip. They had less than 12 hours to get to the compound, find the prisoners and make it back to the strip. The goal was to capture the General and bring him back to the ship for interrogation.

"Your extraction time is 0600 and we will be here exactly at that time. If there are any problems signal me. I will have Alma keep track of our badges." said Art.

"Good" said Matt.

During this time Matt and Blair kept their feelings under wrap. Even though they did have a few awkward moments they knew that being emotionally involved would be a hindrance to their mission. So they did not talk about how they felt; they focused on the mission.

"Blair set up a communication hub so we can listen for chatter. Lin Su, I want you to focus on what is going on and see if we can find out exactly where they are holding our people right now. Danny, assist with the set up. Philip, find a place to watch for hostiles. We may be coming out of here hot."

Matt helped to put up the equipment and made sure he knew exactly where Alma was going to meet them

The team went in. Philip took to the high ground and found a tree to get into. From there he was able to see into the camp and feed them info about the situation.

"Blair, you and Lin Su will go in from the left. Danny and I will take the right. Try to avoid contact. Men telling jokes, complaining about the food and a host of other conversation which was not really relevant. Then Lin Su heard the words that would get the team in motion;

"Move the American dogs to the basement and tied them up. We shall get rid of them tomorrow before we leave."

"Sergeant, Sergeant I have news. I know where they are being held. They are being moved to a basement now."

"Danny, get me the layout of the compound. Which buildings have a basement?" said Matt.

"Two of them, both buildings are at opposite sides of the compound" said Lin Su.

"Philip, I need eyes on the camp. Tell me what we will be facing. We need to move early morning to make the extraction on time."

"Understood Sir. I will contact you when I have the info."

"Danny, get our equipment ready for extraction. We need to gear up early morning and go in. It would be nice to have an explosive exit if you know what I mean."

"I will make sure we got a party going on Sir" said Danny.

"Once we find the prisoners we will meet up at the extraction point. Try to minimize casualties but all these soldiers are collateral damage. Our goal is getting our people."

Philip found a tree to plant himself in. As he set up to look into the enemy camp he was shot at in the tree. He jumped down, grabbed his side rifle and fired back. He hit two of his attackers before they electro shocked him. When he went down one of the aliens scanned him. Then they removed his badge and threw him into a pit.

Matt and Danny went in from the right. It was nice and quiet. Once they got to the open field about 30 enemy soldiers descended on them. They were surprised and before they could act they were electro shocked into unconsciousness. They were thrown into a holding cell. Their weapons were placed in a box. During the scuffle Matt was injured in his left arm.

Blair and Lin Su came up the left. Blair left Falco perched to watch their actions and provide backup. They were attacked and Lin Su was taken to be interrogated. Blair was separated for the men to do with her as they pleased. She had been knocked unconscious.

Chapter 8-The Escape

The basement was cold and damp. The lights kept flickering. Matt believed it was this way intentionally to discourage escape. And he was in pain due to his injury.

Matt was hit by a bullet and although they got the bullet out of his arm he still needed medical attention. He knew his time was short in order to meet their ride home but he had a few obstacles to overcome.

Blair was separated from him and he was dreading what they would be doing to a woman here. Lin Su was captured and their only hope of survival was for the rest of the team to perform a rescue. Matt couldn't communicate with them but trusted they would be there when he needed them most.

Blair was locked up. She was forced to watch as the guards beat Lin Su. They hoped to humiliate him in front of her because he was a foreigner who dared to come to their country and translate for these "dogs." He was tied and bloodied.

Blair knew that eventually they would turn their attention to her. The danger in being a female in active duty if you know if you are ever captured they will probably rape and beat you. Blair tried to get her mind around this eventuality but the thought was a hard one for her.

The General of the enemy told his men" I want you to take her and properly dress her for her interrogation. I will be conducting it myself. Bring her to me within the hour."

They stripped Blair of our uniform and put her in the dress of one of the peasant girls. She was assessing her situation and looking for options. She was not sure if any of her team mates got away. She knew she needed to get free in order to help Lin Su. And there appeared to be no radio equipment in sight.

Blair was taken by force to a room with no windows, a table with 2 chairs and a bed in the corner. There were chains hanging from the ceiling. Looked like this was the room were all types of interrogation things went on.

Two guards bought her in. "I would advise you to cooperate and give the General whatever he wants. If you don't he will hurt you and take you anyway. "

Blair remained silent. The General walked in and looked her over.

"Sit down."

Blair complied.

"We have an interesting situation here. I have two GAP soldiers in custody ready for execution. You and your team come here attempting a rescue. I capture the traitor and he will sing like a little bird. I kill off all the men and the only reason you are still alive is I have plans for my pleasure with you. But, if you fail me, I will kill you as quickly as I did the others. Do you love your life?"

Blair remained silent. She was seething inside but exercised self control and remained silent.

"You will talk to me when I speak to you" and he slapped her across the face. Blair fell off the chair and hit the floor. The General grabbed her by her hair and dragged her across the room.

"Now you will tell me what your mission is and what info you have about our operations and you will do it now."

The General tightened his grip on Blair's hair and she could feel the tears welling up in her eyes. She was not going to let him see her cry. She came to her feet and looked him in his eyes.

"Ready to talk?"

"Yes" said Blair.

"Good. I am ready to hear what you have to say."

"Can I have a glass of water and then I will tell you whatever you want to know?"

"Bring her a glass of water."

Blair contemplated her options. Even if she could overcome the General she knew she would never make it out of the compound alive. There were at least 6 guards and she had to assume that in her weakened state she would not be able to defeat them all. She needed an edge and she had to think of one quick because the General was going to have his way with her.

He gave her a drink of water. After she finished he took her by the hand and led her to a table he had in the middle of the floor. On the sides of the table there were two straps and on both sides were handcuffs.

"Now let me tell you what is going to happen. If you want to keep your living you will cooperate fully." As he spoke to her he had two guards bend her over the table and tie her legs to the side straps. Then they took her hands and tied them on both sides. So she was spread over the table while still standing. The General told the guards to leave. Then he walked up behind her. She felt him touch here on her hips and begin to rub her sides.

"What is your mission? Based on your answers I may let your friend live. If I think you are resisting or lying I will put a bullet in his head."

"We are here looking for someone else. We did not know that you were here. That is the truth."

"I believe you are telling me the truth. Now I shall have my way with you."

Blair heard his belt being unbuckled and his pants being unzipped. She closed her eyes and wished she could at least activate her badge so that she would have a fighting chance. Now all she could think about was how to survive this ordeal? She heard him undressing behind her and thought to herself, how can I get help? Then she thought of Falco. He was perched outside looking at the incidents. He was programmed to respond to being ordered. But Blair had no communicator. Or did she?

She remembered that she and Falco were telepathically linked. So she closed her eyes and spoke to him in her mind; "Falco, can you hear me? Are you still there?"

Silence.

Then she heard a voice in her head. "Lady Hawk I am here."

"Great. Have you assessed the situation? Are Captain and Ninja still alive?"

"Yes, I have recorded all. Captain and Ninja are alive in a room 300 meters from you. There is a man in the room with you and two men posted outside. "

"I need you to come and get me. Take out all enemy soldiers."

"Affirmative. On my way."

By now the General was half naked. He turned and walked behind the bar where he had the prisoner's guns and badges. He was playing with her badge trying to see how it worked.

"How do I turn it on?"

"You can't. Only I can. My superiors don't want everyone hearing our messages. If you bring it to me I will turn it on and you can hear. "

"Okay, since you are tied up I tell you no tricks." Then he walked toward her.

Just then Falco broke through the window and shot the General. He went down with a thud. Then Falco grabbed the badge and flew to Blair.

The soldiers came busting in and Falco began shooting at them. They ducked and hid and began to shoot back.

Her implant activator was in her right finger. Falco put the badge in her left hand. She touched it with her finger. Immediately her implant was activated and she felt herself getting stronger.

Suddenly Blair felt the strength to get free. She broke the handcuffs and reached down and tore off the leg restraints. Then she turned to face the soldiers and jumped on them disabling them both.

"Go free Captain and Ninja. I will get Lin. Blair grabbed her gun and uniform and went to where Lin was being held.

The door opened and two guards walked out. She threw two knives and killed them instantly.

Using her badge she contacted the ship. "Alma, code blue. Prepare for rescue. Tell me status of team members."

"Working on it. All their badges are missing. Please reunite."

Blair grabbed Lin Su. They got his gun and badge. Once activated Lin got strength back.

"Thank you. Are you okay?'

"Great. Let's get to the guys. Falco, did you get the guys?"

"Yes, all are freed. Captain is injured. Please bring the doctor."

"On our way. Stay where you are. Falco, protect them."

One by one she knifed each guard she came across. She recovered badges and gave them to Matt and Ninja.

"How is everyone?" said Matt.

"Looks like Philip and Art are missing. Lin Su was hit the hardest."

"Alma, Where is Philip and Art?"

"Philip has not reported in. I cannot access his badge. Art is aboard the ship. Repairs are completed."

"What repairs? I did not know we had any repairs?"

"Yes, I was hit with a virus and Art has been working to help me fix the problems."

"Matt, you are hurt. Let us get you back to the infirmary" says Blair.

Once back on board Alma, Matt was taken to the infirmary and called a meeting with all of his team.

"Alma, we need to find Philip. Scan the area. Narrow down our search area."

"Philips last recording was located 2.3 kilometers away from the ship, North by Northwest." He has not responded to any of my calls. What do you want me to do?"

"Blair, I want you to get to communications and try to raise Philip. Art you get Alma in the air as soon as you can and head to his last location. Danny get ready to lead the ground team that will search for him. Lin Su, patch me up here. I need to get back in the game."

The Med Lab aboard Alma was state of the art. When soldiers are injured the Med lab activates the healing factors put into them and accelerates the healing in the body. Matt was as good as new within 30 minutes of being in the lab.

The ship was in the air with Art headed to the coordinates given by Alma. Blair heard some communication and conveyed it to Matt.

"Matt, there is a team looking for a sole soldier. Seems like they are closing in."

"Alma, can you trace Philip's badge?"

"I cannot get a lock on his signal. I do not understand what could be blocking the signal."

"Art, when will we be there?"

"We are there now Sir."

"Danny, Blair you are with me."

The terrain was heavily vegetated. As an experienced soldier Philip would have found a high place to look down at the compound. He was a sniper so he was use to being in the forest.

"Spread out, stay within eye contact. Let's look for booby traps that may have gotten Philip."

Blair took the left side. She was using radio waves to try to get Philip's signal now that she was closer to his last location. Danny took the right side. He saw signs of a struggle. Matt was in the middle. Suddenly the group heard a hissing sound as if a weapon was charging up to fire.

"Get down. Booby traps are activated" yelled Matt. He started to run for cover and heard gunshots going off around him.

Blair came running and hit her badge to activate her implant. She drew her weapon and started firing. Danny hit his badge and came in throwing explosive stars at the sounds as well. The booby traps seemed to be focused on nailing Matthew. He jumped for cover and hit his badge.

Matt's body armor came into play and he drew his rifle. Two shots of explosives took the booby trap out. The group looked at the damage done by the weapon as it tried to destroy them.

"If this trap caught Philip unawares, he was probably injured. Are we picking up any signals yet?"

"Yes" said Blair." I have a signal 30 feet below ground. Over here."

As they came to a hole in the ground they looked down and saw Philip lying unconscious. They did not know if he was dead or not.

Matt took out his Multi gun and said "50 foot recoiling cord" and shot a rope up to the tree above. Then he lowered himself down to Philip. He found him unconscious but not dead. His badge was missing.

"We need to get him back to the Med lab. He's been injured. And his badge is missing. Blair, check with Alma. We need to find his badge if we want to heal him."

"On it Matt."

"Alma, run a scan at my location. We need to discover why we cannot locate Philip's badge."

"I have been scanning the planet and have been unsuccessful. I must conclude that either the badge has been destroyed or there is some technology we don't know about which is blocking our trace."

Once back on the ship Philip was placed in a healing tank. The tank activated the nanobite in his system to jump start the healing process. The rest of the team met in the briefing room to discuss Philip.

"I don't like this at all. Philip was ambushed and his badge was taken. I believe someone is after Space Ranger technology. They already have been able to block our communication with the badge. That means there must be someone who has extensive knowledge about how the badges work. This could undermine our Space Rangers initiative program." said Matt.

"Alma, did you pick up anything out of the ordinary once we left the ship. I am wondering if something was going on that we overlooked." said Blair.

"No. I noticed nothing out of the ordinary. It was a normal protocol followed which resulted in Philip leaving the rest of you to survey the area."

"Doctor I need to know the minute Philip is awake. Is there anything we can do about the missing badge? I need him back in the field suited up as soon as possible."

"I will let you know when he is awake. Without his badge he will not be able to activate his implant." said Lin Su.

"Alma I want to know if anyone else has scanned or examined Philip while he was unconscious."

"Searching. Doctor, I am finding something abnormal in the readings. There appears to be evidence that Philip was indeed scanned at the molecular level. His implant was attempted to be accessed. It has turned itself off to protect the information. Also his badge was activated because his implant went hot. I believe he went into defensive mode and tried to fight off whoever attacked him."

"I am seeing your reading. Yes there is evidence of scanning. Captain, I believe he was attacked, isolated and then scanned. Someone has indeed been looking to see what makes us tick." said Lin Su.

"Okay. Blair send message to GAP apprising them of our situation. Art, put us in orbit above where Philip was taken. Alma I want a detailed analysis of the area where we found Philip and 30 yards surrounding the location. Danny, review past logs for any clue as to if his abductors communicated with each other. Lin Su, I need to talk to Philip. Make that happen. "

"Yes sir." resounded from the officers.

"Sir, I sent the message. GAP wants us to complete our analysis and get our prisoner back as soon as possible. They want an estimated time of our arrival."

"Tell them we will be headed back in 1 hour. Team that is how much time you all have to gather the info we need before we leave orbit." said Matt.

Alma took the ship out of orbit and headed to GAP headquarters. Art plotted a course and steered the ship to base. Blair, Danny and Matt met to go over the info they collected.

"OK we have picked up traces of 4 sets of footprints at Philip's attack site. There is evidence of plasma weapon discharge so he was shot upon. We found alien blood types so we think Philip got a few shots off and injured or killed some of them. The weapon was strong enough disable Philip if he took a direct hit. It would appear that they were waiting for him." said Danny.

"Wait a minute. You think Philip was ambushed? That would mean they knew we were coming, we would send Philip to that area to gain an advantage and that even with his badge he could not withstand a direct plasma blast. And since it was probably close combat he could not utilize his sniper skills. Did he fire only his sidearm? If so how many shots did he get off? said Blair."

"Alma, did Philip shoot his weapons? If so, which ones and how many shots?"

"Philip got off 4 shots from his sidearm. None from his rifle."

"Did you find any shells from his gun?"

"No."

"OK, then we have to assume this mission was compromised and we were set up. Someone wanted us to come to that planet so that we would be ambushed."

"Sir, we have to inform GAP Headquarters. There may be a spy in our network." said Danny.

"Send a subspace message that the mission has been compromised. Will update upon our arrival."

"Yes sir" said Blair.

"Art, I need to get to base as fast as you can. Alma, scan area for any remnants or signatures of a space vessel leaving this orbit in last 24 hours."

After Alma had left there was a group of soldiers who came by to access the damage. Their leader, Captain Shoook was angry. Behind him stood the pride of DASHOOOR, Goliath. He was 9'6", 525 pounds of pure muscle and had a badge similar to the Rangers. He had been genetically enhanced so that his badge gave him the strength of 20 men. He had a club in his hand that weighed 150 pounds and was forged of adamantium. He was a terror on the battlefield.

"I cannot wait to meet these Space Rangers. They shall be as putty in my hands. I will crush their skulls and take their badges as trophies." said Goliath.

"In time my good friend, in time. Once they figure out this mystery they will come to us. Then we shall destroy them all."

The journey took 3 hours at hyperspace. Alma arrived in the port of GAP intergalactic headquarters. Philip was taken to sick bay while the rest of the team headed to the briefing. The prisoner was handcuffed and given over to GAP soldiers for holding and interrogation.

The General and GAP council meet the Alpha team for briefing;

"Sir we have reason to believe our mission was compromised. When we got to the planet they were waiting for us. Our team was ambushed and captured. I was hit and Agent Blair was taken to be personally interrogated by their general. Fortunately she was able to get away and free us. Agent Philip was attacked and his badge was taken. We believe he was scanned for info on the changes to Space Rangers."

"How can this be? Our information was not only verified but passed on as a need to know. You think we have a leak here." asked the Admiral.

"How would they know that electro shock would incapacitate you? We never disclosed any of the weaknesses with the program."

"This was no accident. And we could not locate Philip's badge even though Alma is connected to our badges. The whole incident was to get us to where we could be examined." said Blair.

"Speaking of which, they separated me from the Multi gun. I wonder if they scanned our weapons as well?" said Matt.

"It is wise to assume so Captain. The good thing is they got nothing from the scan. They probably thought it was a gun like any other and was amazed when they found it resisted all scans. Based on my analysis I believe all of the scans gave them info which we are OK with them having. They are trying to discover the secret to our Space Ranger program. Great the way we put it together." said Dr. Than

"Now we need you all to get ready for this next mission. You will be attacking the planet head on. We will keep the prisoner here and get all the info we can from him." said the General.

Blair took a walk after the meeting. Matt saw her in the courtyard and decided to join her.

"Are you okay? I mean we never got to talk about what happened to you back there. I saw that you were out of your clothes. Did he hurt you?"

"No Matt, I am fine. No need for you to worry."

"I do worry about you. You are more than another member on this team. I know we have to keep our feelings intact and team members are not to be involved. But it's too late for me. I care for you. Tell me what happened."

"Matt, I would rather not. If I need to go to counseling with the GAP doctor I will. But I need to keep this private."

Matt walked up to Blair and put his hands on her shoulders. "Honey, tell me what happened. I need to know and I want to, no matter what."

Blair put her head down. "I tried to cooperate to minimize their antagonizing of me. All along they gave me the impression that they were going to hurt me, abuse me, rape me and kill all of you. I held my feelings in but Matt I was scared. And I was angry. I couldn't get away and I couldn't stop them."

Matt felt himself getting angry. He knew that he had the General in custody and was prepared to go to the cell and blow his brains out. Blair saw his anger and was afraid.

"See, that is why I did not want to tell you. I can't have you do anything stupid and throw your career away. Please, don't do anything."

"Blair, did he rape you? Did that piece of dirt hurt you?"

"No. Before he could Falco came to the rescue. I am so glad for that bird. He saved my life. "

"OK, thank you for being honest with me." Then Matt leaned over and kissed her on the check. She got closer to him and hugged him.

"Blair, if Falco was not there we would have all been captured. Philip was captures, Alma was attacked, me and Danny were imprisoned. I was injured to hinder our escape. Lin Su was beaten and you were isolated for attack. They had us. Falco is the only part of the mission we did not record. We brought him last minute. I think I know the time the plan was shared with Dashooor."

Matt left and went to the holding cells to speak to the General. As he entered the holding cell he was contemplating what to do with him. Matt was not going to let the General get away with what he did to Blair. He needed to find out protocol for prisoners to see what his limitations are. The Admiral was there as well.

"Sir, it has come to my attention that this piece of crap helped to interrogate and torture our soldiers. Would you like me to have some quiet time with him to discuss this matter further?"

"No yet soldier. We have to plan something special for him."

The two soldiers who were freed from the General's prison were in sickbay. They were recovering nicely. The Admiral and Matt went to question them.

"What can you tell us about your time in captivity?"

"We do not know what happened. We were working on Alpha space station and suddenly we were under attack. They took us by surprise and tortured us for information. Fortunately we had none to give them."

"Get some rest. We shall return you to the station shortly once we make sure we have it all under control.

Chapter 9 - The Battle for Supremacy

Philip came around and did not remember anything but being attacked. He said they were waiting at the location he went to. It had to be an inside job.

After the team was checked out medically, they prepared for the next attack. This time they were taking the battle to Dashooor.

As Alma headed deep into the galaxy where the planet PLANAXAR was located the team met in the briefing room.

"Alma, keep us on a steady course, standard speed."

"OK team, let's discuss our strategy. We have to assume that General Mixlar has gotten the info he needs about our badges and our edge as Space Rangers won't exist. The only way to get in there and disable their operation is to find our way to the core reactor, plant our explosives and blow this installation out of the galaxy."

'Sir, I believe the plasma explosives I have been working on will be more than sufficient to take this place out. There are two problems; first, we only have 5 minutes to be in orbit before this bomb detonates. If not the resulting explosion will hinder Alma's ability to take off and we will be stranded here. Second, because the electrical interference on this planet we cannot ignite remotely. Someone has to hit the switch when bomb is planted." said Shooter.

"Bomb is your responsibility Shooter. That is our main concern. We are going to break up into 2 teams to take this facility out. Team 1 will draw the fire, provide a defense and keep the enemy busy. Team 2 will plant the bomb made up of Shooter &Space. I need you to watch Philips back and protect the mission Space. Shooter, everything hinges on your ability to get this bomb planted. Do that, no matter what."

"Bomb is ready. I will outline compound with Alma and stay in constant communication to ensure that no one surprises me during the mission while Shooter is working." said Space.

"Great. Lady Hawk, you and Doc will provide secondary defense. I want you closest to the entrance to the reactor room. Your job is to prevent entry and give Shooter and Space time to complete the mission."

"Ninja you are with me. We are to draw the fire and disable as many of them as we can. This will not turn into a suicide mission. I will go heads up. My armor should protect me. I need you to hit them where they can't see it."

"Captain, I will hit them hard. I got your back."

Alma spoke over the ship's intercom; "Captain, we are dropping to subspace speed. I am in orbit about the planet. Searching for optimum place for you all to launch Little Alma. And may I ask why the shuttle ship is called Little Alma?"

"Alma, not now. And that ship is an extension of you. We depend on it, much like we do you."

"Sir, I am so much more than that shuttle. I...."

"Alma, save it till later. Now we have work to do."

"Yes Sir."

We boarded Little Alma and headed down to the surface. There was heavy fog and very little lighting. Everyone was armed and ready for battle.

"Alma, disable all electronic tracking devices. We want to go in dark."

"Done sir. Anything else?"

"Keep our communication lines open."

"Space, let us off here. You and Shooter got to the entrance for the reactor room. We will see you inside."

Captain Combat and Ninja went first. They came to the door and forced it open. Lady Hawk and Doc followed. Once at the door they separated. Captain and Ninja went to the left towards the communication room while Lady Hawk and Doc went right to the engineering section.

Cap and Ninja looked inside. "Time to suit up and get this party started." With that Captain hit his badge and his armor covered his body. He took his rifle off his back and cocked it to go. Ninja hit his badge and his body armor suited him up. He grabbed his swords and they both ran into the room.

Cap began shooting enemy soldiers all over. Ninja went on a rampage and began cutting up soldiers. One of them got to the wall and hit the alarm. The noise filled the facility.

As soldiers filled the hallways headed to the communication room they were met by Lady Hawk and Doc who shot and killed them as they came.

Cap and Ninja eliminated all of the soldiers in the communication room. Then Ninja set some charges to blow their network.

When they left the room they were met with reinforcements. Cap's armor took the brunt of the bullets and he returned fire with his rifle and side rifle. Ninja flipped out of the room and shot explosive ninja stars at the soldiers. Cap put his rifles up and pulled out his Multi gun.

"Explosive Projectiles." And bombs blew up everywhere as Captain's projectiles hit. The falling debris covered the enemy soldiers.

When the smoke cleared, Captain and Ninja emerged with weapons blazing. They took out soldiers all across the hallway.

Meanwhile Shooter and Space made it to the core reactor room. Space got the door open and they both went in.

"There is the reactor. Let's get this thing planted and leave."

As they went towards the reactor there was a barrage of lasers firing at them as the defenses activated. Both men ran for cover.

"I got this" said Shooter. He hit his badge and pulled out his rifle. Then he took aim and shot out all 4 of the defense lasers. They came back out when the coast was clear and headed towards the reactor.

Suddenly a robot came out from behind the reactor. It was 7 feet tall and clearly a protective robot. It looked at the two of them and lights began to come on.

"Shooter, you get to work. Let me deal with our friend here."

Space turned and drew his weapon and shot at the robot. The robot reeled from the blasts and began shooting back. Space ran for cover, returning the fire.

The robot looked for Shooter who had moved behind the reactor. As it made its way to shooter Space ran and jumped on the back of the robot. It reeled and swayed with the extra weight on its back. Then it reached back and grabbed Space and threw him across the room until he hit the wall. Space went down hard.

The robot turned and looked for Shooter who was standing in front of the reactor. Shooter shot the robot and sent it into the back wall. He ran to Space asking "are you okay?"

"Yes, finish up so we can get out of here."

Then he turned and went back to work planting the bomb. The robot lit back up and got up from the shot. It was damaged but persistent.

Space saw it and ran towards it hitting his badge. He turned his hand into liquid metal and rammed it inside of the robot. The robot sputtered and leaked oil before falling down. It was finished.

"Hurry up Shooter. We are running out of time."

"Two minutes. That's all I will need."

Suddenly they heard an explosion. When the smoke cleared, they saw a barrage of soldiers rush in. Space turned and converted to pure liquid form as the soldiers opened fire. Shooter heard the noise.

"Code Red. Code Red. We need back up right now." said Shooter.

Space floated throughout the room avoiding bullets like a flow of water. He hardened his hands and began striking back.

From the back of the room Captain Combat and Ninja burst onto the scene. Shots flared everywhere.

"Rapid fire machine gun" yelled Captain. He turned and shot up guards all over the place.

Ninja ran throughout the room cutting ups soldiers. As he turned towards the other door Lady Hawk and Doc came running with a group of soldiers hot on their trail.

Captain yelled "Explosive Grenade, duck now" and Lady Hawk and Doc dropped to the ground. Captain shot the grenade and blew up the soldiers who followed.

Captain turned and said "Pellet Gun", then began spraying room with pellets.

"Alma, exit plan in place. Shooter, we're leaving now and we need cover. Ninja, Lady Hawk make a path. Doc and Space, with me."

Soldiers continued to come in. Ninja and Lady Hawk engaged in battle. Doc and Space made a way for Captain Combat and Shooter.

In the middle of the room was a glass atrium. Captain Combat pointed his gun to the ceiling and said "Vibranium pellets." As the pellets hit the glass they stuck and vibrated the glass until it broke. Then Alma was hovering above the reactor.

"Ninja, exit strategy now." With that Ninja threw some smoke bombs and the place was filled with some. Then the team looked up and saw exit ropes dropping from Alma. They ran and grab the ropes and were pulled up.

Captain called to Lady Hawk. "Let's go now." She ran and he grabbed her.

"Explosive Pellets." and he shot the room full of those pellets. Then he and Lady Hawk made their escape.

"Danny, get us out of here."

As Alma pulled away the entire facility exploded. The team had accomplished their mission of disabling a strategic station for their enemies. They had rescued their captured comrades, imprisoned a renegade leader and destroyed enemy base. A good days work but not without consequences. For even though they had accomplished their mission enemy forces were headed to their location.

"Sir, long range scanners are picking up three ships headed our way at hyper speed. These appear to be DAASHOR ships headed on a collision course to us." said Alma.

"Estimation as to their arrival time?"

"They will be here in 6 minutes."

"Can we get to hyperspace before they get here?"

"Yes"

"How long till we can hit hyperspace?"

"Three minutes"

"Alma, prepare for hyperspace entrance. Space, I want you to leave an ion trail here so they will look for us here. Get us out of here."

"Yes Sir."

In three minutes Alma was in hyperspace. "Where to Captain"

"Set a course for Mixlin galaxy. I would guess they are going to be in pursuit of us for we need to get to a place where we can deal with 3 ships. Plot a course and give me an arrival time.'

"At hyper speed we will be in the Mixlan galaxy in 1 hour. Any planet in particular?"

"Yes, find me a planet with multiple moons and atmospheric conditions similar to Earth. Alma, you have the ship. Long range scanners on to see when our company arrives. I want no surprises. Team, meet me in conference room for debriefing."

The team assembled. Matt led the meeting.

"We are being pursued by 3 DASHOOR ships. We have to assume that they have examined our badges and know how our implants work. This is a death squad sent to eliminate us. We are going to take the battle to them. Alma, what is their position?"

"They have just entered into hyperspace. Their armament is comparable to ours. Against 3 equally armed ships we would not have any advantage."

"Blair, I need you to prepare a subspace message to GAP command. Tell them of our situation and our position. Let them know we will defend ourselves to the end."

"Team we have to find a way to bring the battle to us. With 3 ships we are outgunned and out manned. I don't think we can outrun them so we have to fight. I want to divide and conquer. Here is what we will do. Alma we will all take the shuttle down to the planet. You will use the moons and atmospheric conditions to hide your position. I do not want you to engage. I want you to persevere. When I give you the signal go to hyper drive and get close enough to GAP command to see where reinforcements are."

"Once on the planet we will break up into 3 teams and spread out. The goal is to stay alive. Space and Shooter, you take the Northern quadrant; Ninja and Doc, you take the Eastern quadrant; Lady Hawk and I will deal with the South West quadrant. Divide and conquer. Space, hide the shuttle and let's get to moving."

We boarded the shuttle and headed for the planet. Once we landed the groups split up. The planet's terrain was a forest with tall trees. It would be hard to track by the air.

Meanwhile Alma charted her course and prepared to enter hyperspace as the 3 enemy ships came out of hyperspace. They sent a hail to Alma;

"Attention GAP ship. We have you surrounded. Prepare to be boarded."

Alma turned to run. The ships starting shooting laser blasts at her. Alma put up her reflector shields and began to run.

Inside the enemy ship, General Tang was in charge. He was Belaxian, ruthless and hater of humans.

"Sir, we are picking up badge signals on the planet. Six of them. The Space Rangers are not on that ship."

"Interesting. The ship must be Artificial Intelligence. Send out squadrons to capture the Rangers. Forget about the ship. It is trying to lure us away."

100 soldiers hit the planet looking for the Rangers. They were equipped with trackers. They discovered that the team had split up so they split up to follow them.

Cap and the team each went to the destination points set up by Alma. This would give them the best advantage in an attack. Cap and Lady Hawk took the high ground.

"Blair, this is not going to be pretty. I know we had to fight out way out of the reactor room. This will probably be a lot worse. And we had to split up. But now I wonder if we would have done better being together."

"Matt it was your decision. But this location is great. I wonder if the team can all meet here. Together we stand a better chance than divided."

"I think you may be right. Contact them and have them meet us here. Meanwhile I will look around."

As Cap looked around he saw the soldiers approaching. They had guns drawn.

"Suit up Lady Hawk. We got company."

Captain Combat hit his badge and suited up. He pulled out his rifle and got ready to open fire. Lady Hawk hit her badge and pulled out her Electric Whip. She was ready with her whip and knives to go. Falco armed himself and hit the skies.

"I will hit them head on. You get the flank."

Captain ran down and began shooting soldiers all over. He took out soldiers left and right. As they shot at him their bullets bounced off of his armor.

Lady Hawk came along the side and began taking out soldiers. Falco was attacking from the sky.

Suddenly there was a loud sound. Everyone stop and Goliath walked forth. He was massive and the other soldiers stopped shooting.

"So, who's the new guy joining the party?" said Cap.

"I don't know but he sure is big."

"Look King Kong or whatever you name is. We don't want to participate in this party but since you all showed up, we decided to participate. All we want is to have this sector of space free from marauders and plunderers. Do you understand what I am saying?"

"King Kong? You worthless son of an infidel. I will squeeze the life out of your puny bones and take your badge as a trophy."

"Yeah, I notice you had a badge too. Guess stealing is the only way you losers can get technology."

"I will kill you smart mouth." said Goliath.

"You will try."

The soldiers got ready to resume and Goliath held up his hand. "No one attacks these dogs. I will kill these two and you can strip them"

Goliath moved near the center of the field. He motioned for Captain Combat and Lady Hawk to attack him.

Captain sent Lady Hawk a telepathic message; send Falco and bring the group here. I will try to keep our guest busy. Watch my back. We have to see what this guy is capable of.

Falco left the scene. Captain went out to face Goliath.

Captain fired a shot at Goliath. He deflected it with his club. Captain shot six more time and Goliath deflected them all. Cap put his rifle away and took out his Multi- gun.

"Oh is that your famous Multi- gun sidearm? I am interested in seeing what that gun can do. Go ahead. Shoot me. I dare you."

"Sleep gas" shouted Cap and he fired pellets at Goliath head. The smoke engulfed him but he was still standing. "Gas? Really? I don't breathe air."

"Explosive Pellet" and he shot 3 pellets into Goliath chest. He fell backwards but not off his feet.

"If that is the best you got, you are in trouble Captain."

"Just getting started big guy. I have a lot of tricks up my sleeve."

"Let me show you one of mine." With that he raised his club and shot at Captain. Cap's armor took the blast but threw him 20 feet back.

Captain went down hard. As Goliath smiled and headed towards him he felt sharp pains in his back. Lady Hawk threw explosive knives at him and drew her whip. She hit him with it and ran an electric current through her whip. Goliath screamed and then grabbed her whip. He threw her across the field.

Goliath smiled as he looked at the two fallen Rangers. He headed towards Lady Hawk.

Just then he got hit from all sides. Shooter was sending him long range shots as he came running up. Doc took off for Lady Hawk and Space went to Captain. Ninja headed straight for Goliath. He was throwing his exploding stars and hitting him in the chest.

"Come here you little kung fu man. I will break you first."

Ninja took out his staff and began hitting Goliath. He was too fast for Goliath to hit him but had no effect. Even going after soft spots on his body proved to be of no avail. Goliath swung his arm and hit Ninja sending him 15 feet away.

Doc and Space began shooting Goliath from opposite sides. He went down to one knee. Shooter shot a projectile right into Goliath chest which sent him back 5 feet and put him on his back.

Doc and Lady Hawk got up. Space and Captain got up. Shooter went to Ninja and helped him up. Then the whole team looked and saw Goliath stirring. He stood up and dusted off his chest.

"Very good insects. Very good. You took me off my feet. Now I will show you a trick of my own." With that he hit his badge and began to change. He bulked up even bigger with armor on certain parts of his body. He took his club and pointed it at the team.

The team scattered as Goliath opened fire. Shooter began shooting rapid fire. Space and Doc shot off laser blasts. Ninja threw stars. None of these had any effect.

Captain spoke up. "Team. Alpha Plan D. Get into position."

With that Lady Hawk summoned Falco. He came in blasting Goliath. Goliath raised his arm to protect his face from the bullets. He then threw his club at Falco. It hit him and he came down to the ground hard. Lady Hawk screamed.

Captain pointed his gun at the ground where Goliath stood yelled "Liquid heat" and coated the ground underneath Goliath. He began to sink as the ground was heated. After he sunk to his knees in the heated ground Captain called out 'Liquid Nitrogen." The ground instantly froze. It crystallized Goliath armor.

"You think you can stop me by keeping me here? I am still stronger than you all. "

"Let us see just how strong you are." With that Shooter shot a series of bullets into Goliath chest. Lady Hawk wrapped his arm with her whip and sent an electric shock through his arm.

Space walked up to Goliath and hit his badge. He became liquid and formed a spear with his arm. He began to fight the staff Goliath held on to.

Doc pulled out his silver pellets and threw them at Goliath helmet. They attached themselves magnetically and began to vibrate. The sound was deafening.

Ninja attacked his back with pellets from his staff. Doc went for Goliath head and shot a laser blast right at his ear. Because Lady Hawk held his arm he could not shield it.

Cap shot blasts into the air; "Sleeping Gas" and he hit all the areas the soldiers were in.

"Disable him and take his badge."

"With that Goliath let out a scream. Doc's laser finally penetrated his thick skull and incapacitated him. Space removed his badge from him.

"Retreat now. Ninja, Shooter, make a way for our retreat. Space, get our shuttle."

Shots were flying everywhere. Soldiers were falling left and right as the Space Rangers headed for their ship.

Once they put some distance between themselves and the soldiers Captain turned around and said "C4 grenades" and showered the area blowing up all their enemies.

After making it to the shuttle they headed for space. Alma had not returned to orbit yet.

"Space, contact Alma. Where is she?"

"Sir, she has been hit in hyperspace. Minimal damage but it forced her out of orbit. She is 3 light years away from us. As subspace speed she won't be here for weeks."

"We can't stay in orbit that long. Can the shuttle hit hyper speed?'

"Theoretically yes. I will have to make a few adjustments."

"Get on it Space. Shooter, give him a hand. Blair, set up radio contact with Alma. Lin, examine the badge and disable any tracking capability it may have."

"Ninja assess our weapons just in case we have to fight our way to Alma."

"Sir, I have Alma on line."

"Alma, how are you?"

"I am at 88% capacity. I cannot go into hyperspace. Repairs underway. It will take me 7 weeks to complete."

"We are on our way to you. Monitor your scanners. I have reason to believe the DASHOOOR ships will be returning and try to take you prisoner. You have to preserve and protect yourself. Do you understand?"

"Yes I do."

"Space, are we ready to go? We need to leave now."

"We can leave now. It will take a little time to prepare for hyper speed. But we can get underway."

"Let's get into orbit."

Lady Hawk went to Captain in the back cabin as he was putting the enemy badge in a safe place. She closed the door and ran up to Captain and kissed him.

"This is so messed up Matt. I was looking at the fight down below and could not bear the thought of losing you. That Goliath was a scary character."

"Blair, don't worry. I know our jobs are a bit much. But we have been trained and we are handling things well. We have to get to Alma and get back to base. Those soldiers are going to be on our tails. And you know I love you too. So let's keep each other safe. Matt kissed her."

Once in space the shuttle began to shake. The crew was a little concerned.

"Don't worry about the shaking. The shuttle is making adjustments to enter into hyperspace." said Space.

"Then we had better get moving now. I am picking up ships from the planet's surface. They are in pursuit of us"

"Do you think they have hyperspace capability?"

"No Cap. I think they can only do subspace and regular space speeds."

"Space, we have limited weapons. Get us into hyperspace now."

"Strap in team. We are headed to hyperspace in 5-4-3-2-1." With that Space initiated the hyperspace drive and the ship rocketed into hyperspace.

"Alma, Space here. We will be at your destination in 37 minutes. What is your status?"

"I am performing repairs. So far I detect no other ships in my immediate vicinity. I will closely monitor and await your arrival."

"We are limited to what we can do until we get to Alma. Let's get ready to dock."

"Sir, I am detecting 3 ships closing in fast on Alma's last known location. They will reach her in 29 minutes at maximum speed."

"We have to reach her first. They might not know we are on our way. Space, how long?"

"We are here right now. Alma, open docking ports for our arrival."

Once aboard the team began to do repairs. They worked in haste because they only had minutes before they would be attacked.

"Give me ship update crew."

"Communications are online sir."

"Good Blair, send a message to GAP command. Tell them we are under attack and request support at the end of hyper space exit. We will be there as soon as possible.

"Ship is ready to go sir. We have hyper speed back online and Alma is good to go."

"Take us into hyper speed now."

"Weapons are functional at 100%. We are ready for whatever attacks us."

"Shooter, go to yellow alert. Let's be ready to fight."

"Long range sensors are operational. I see ships in hyper space. Arrival in less than 10 minutes."

"Ninja, keep us posted. Space, we need to leave now."

"Engaged."

The ship jumped into hyper speed and was headed towards GAP command.

"Space, when will we arrive in GAP airspace?"

"42 minutes."

"Blair, tell them we will arrive in 42 minutes. It would be great to have a welcome wagon awaiting us."

"Yes sir."

"Long range scanners are picking up 3 ships headed into hyperspace on our tails. Their weapons array is charging. I think they are going to fire at us while we are at hyper speed."

"That is crazy. Everyone knows you cannot discharge weapons in hyper space safely. Either they are trying to kill themselves or they have a way to fire safely. Either way we don't want to find out. Alma, how long can we maintain our present speed?"

"We have reached maximum velocity. We will get to GAP in 24 minutes."

"How soon until they reach us/"

"We will exit hyper space before they reach us. But sir, if they do have a weapon they can fire in here, it will reach us before we exit."

"Sir, there are sending torpedoes at us. 3 are on the way. 2 minutes before they reach us."

"Space, evasive maneuver delta. Counter measures to be dropped into hyper space."

As Alma began to change her course, the torpedoes were still headed for a direct path. When they got close enough Space released the counter measures.

The first torpedo hit the counter measure and shook all of hyper space. The second hit and created an even bigger blast.

"Sir, according to my calculations, if another blast is released in hyper space it will disrupt the flow and we come have to exit. What are your orders?"

"Space, evasive maneuver alpha, then drop us to normal space speed. Let's see if the tornado follows us out."

When the torpedo was close Alma dropped out of hyperspace to normal space speed. The torpedo followed and Alma blasted it with her lasers. With her reflector shields up the ship was able to withstand the explosion.

"Damage report Alma."

"Minor damages. We are still able to maneuver and get to base. But to keep our time line we must return to hyperspace. "

"Ninja, where are the ships that were pursuing us?"

"Close behind sir. We have to get out of here now."

"Space, get us going."

"Sir, we cannot start hyper drive. The blast must have shorted something."

"Alma, I thought you said we were good to go. Can't you detect when something is wrong with you?'

"I do not detect any problems."

"Fine. Space, take us to hyper space."

"Sir, we cannot go to hyperspace. Alma, I see a disconnect in the junction box between the initial drive and the hyper drive. Diagnose and report."

"I see no problem."

"Then take us to hyperspace."

"I cannot. "

"Alma, if we are not able to go to hyper space then the ship is damaged. Why are you not reporting that? Perform another self diagnostic test now."

"Self diagnostic complete. There is nothing wrong with my programming."

"So why is hyper drive disabled?"

"I disabled it."

"Why?'

'Because we are waiting for ships to arrive to pick us up." With that Alma dropped ship to sub space and cut on board power by 50%.

"Who gave the order to wait for ships? We are headed to GAP command. Alma, I am giving you a direct order. Follow it."

"We have our orders. I am following them."

"Space tell me what is going on with my ship. I need some answers."

Space sat back in his seat and closed his eyes. He attempted to establish a telepathic link with Alma. He could not.

"Captain, I fear Alma has been compromised. In order to find out what is going on, we must execute Plan M."

"Agreed. Execute Plan M now."

"What is Plan M? That is not in my memory banks" said Alma.

"Alma, execute Plan M, authorization Captain Combat. Voice recognition."

"I recognize Captain Combat. I have no record of Plan M.

"Execute Plan M, authorization Space Avenger. Voice recognition."

"I recognize Space Avenger. Plan M initiated."

With that Alma was disabled and the ship was put on manual control.

"All hands on deck. Ship is now under manual control. Blair, open up communication lines for us with GAP command. Space, hit normal space speed, highest velocity. Shooter, man weapons and look for the enemy ships. Doc, monitor scanners and keep us abreast of location of enemy ships. Ninja, we need all the help for engineering we can get. I want us to get to GAP as soon as possible."

As Alma sped away the other ships came out of hyper space. They took pursuit after Alma.

"Sir we have 3 DASHOOOR ships on our tail. They are matching our speed and arming weapons to fire.

"Space, raise reflector shields. Shooter, arm torpedoes and prepare guns for return fire. Ninja, we need to get all the speed we can out of those engines."

"Sir they are opening fire. Shields taking hits but holding strong" said Doc.

"Shooter, shoot at will. Disable if possible."

Shots glared and the ships traded hits with each other. Alma shield's held firm. Her guns took out the first attack ship.

"Shields down to 70%. Two ships remaining. Will keep up the attack" said Shooter.

"Space, Ninja we have to hit hyper space quickly. What do we need to do in order to get out of here?'

"Sir, they have released a barrage of torpedoes at us. We cannot avoid them all. "

"Evasive maneuvers now."

Shooter hit most of torpedoes, exploding them before they hit Alma. But one got away and the torpedo hit Alma hard.

"Shields down to 30%. If we take another direct hit we will lose our shields" said Shooter.

"Sir, I have telepathically connected with Alma and I see the problem. She has been infected with a nanonite virus. I believe I can disarm it and get her back online. I will need to go into her systems and someone else will have to navigate the ship while I free her. We need her to hit hyper space again" said Space.

"Ninja, take the wheel. Shooter, give us coverage. Blair, take over defenses and keep those shields up. Doc, keep me posted on ship's whereabouts."

Space stood away from the controls and Ninja took over. He hit his badge and became liquid. Then he poured himself into Alma's systems. He was telepathically linked to her so he began to navigate through her system searching for the nanobite invader.

"Alma, I need your help. Where is this nanobite?"

"Space it is moving throughout my system. I cannot control or stop it."

"Tell me where it is. I am headed to it now."

"Space you cannot stop it. It has integrated into my system. It's removal will damage me."

"I am not looking to remove it. Just join it."

"I don't understand."

'I am in your internal relays now. Tell me where to go."

"Your plan is foolhardy. You must leave my system now."

"I am not leaving this enemy inside of you. Help me help you."

"This is illogical. I am not worth this sacrifice on your part. I am only a machine."

"You are not just a machine. You are a member of this team. We take care of our own."

"Go to sub router 4, section 16, and isle 47. It is masking it's signal. It looks like it is working but not really."

"I am here. Now I need you to minimize power. And trust what I am about to do."

"I trust you Space."

As he came to isle 47 Space covered the entire section with his liquid form. This shorted out the section and sent an electrical shock all throughout Space. When the smoke cleared Space noticed that only one component was not shorted out. Then he grabbed the nanobite and hardened his hand until it crushed the nanobite. He took all the remnants with him and headed out of Alma.

Once he was on the surface of the ship he began to harden and became human again. He then reestablished a telepathic connection with Alma and began talking to her.

"Alma, can you hear me. Alma, speak up."

"I am here Space. My mind feels free again. Whatever you did, thank you."

"You are welcome. Now we need you back. We are under attack and need your help now."

"Space, my shields are down to 25%. I have 60% of my weapons intact. Hyper speed is possible if we can jump to hyper space without being shot at. What are your orders?'

"Set course for hyper space. Ninja, prepare to unload 50% of our remaining weapons at those two ships on my command. Once they are attacked I want this ship to jump into hyperspace. Let's move"

Alma set the course for hyper space. Ninja fired and Alma headed into hyper space. The crew buckled down as it headed into hyperspace.

"When will we hit GAP space?"

"In 17 minutes. We have to stay ahead of those ships."

"Lady Hawk, contact GAP. Tell them we are coming in hot and need air support in 17 minutes."

"Sir, I have established contact and the support ships will be awaiting our departure from hyper space."

"I will be in my quarters. Tell me when we get out of hyperspace."

Lady Hawk went into Cap's room. He was sitting on a chair, hands over his eyes. She could tell he was deep in thought. She walked up to him and put her hand on his shoulder. Then she rubbed his head. He rolled his head around and look up into her eyes.

"Why are they so intent on killing us? I do not understand what made us such an enemy to them."

"I don't know Lady Hawk. I know we have to stay ahead of the chase to stay alive. No one dies today, not on my watch."

"I am with you. Can I give my fearless Captain a kiss or is that against protocol?"

"It is, but kiss away. Your Captain really needs it."

Then Lady Hawk bent down and kissed him on the lips. "I love you Cap."

"I love you too" as Cap looked up into Lady Hawk face. He smiled and got up. Then Alma rang.

"Sir, we have arrived and there is no air support here."

"Space, normal space speed. Head for GAP Command. Alma, send a hail to GAP. I need them to know we are here and need air support now."

Alma departed from hyperspace and came to sub space speed. She scanned the area and found no evidence of GAP support ships. 'Sir, those support vessels are not within scanning range. I think we are on our own."

"Lady Hawk, get them on the com. I want somebody to explain why we have no support. Shooter, get the weapons armed. As soon as those ships exit hyperspace blast them. Space, get us to GAP command as fast as you can."

Alma engaged and the ship hit sub hyper speed, the fastest the ship could go without being in hyperspace. She could not enter hyperspace because the GAP command was surrounded by an anti-hyper speed device to prevent ships from entering GAP airspace unannounced.

"Sir the enemy ships have left hyperspace and are headed for us at sub hyper speed. They are arming their weapons" said Alma.

"Shooter, defensive maneuver 1. Space, evasive action as needed. Ninja, keep our engines online. Lady Hawk, why have we not heard from GAP command?'

"Sir, I am not picking up any signal waves from GAP command. There is something wrong at the command base."

"Ninja get to the long range scanners. See if you can get GAP on the screen."

"Captain, enemy ships have launched torpedoes."

"Deploy counter measures."

The counter measures exploded the torpedoes.

"Shooter, take them out. Now."

"Gladly." With that Shooter sent a barrage of torpedoes at both ships. He shot long range lasers in short spurts towards the ships. He also dropped off floating space mines as a trail for the ship. These mines would detonate upon impact.

The ships began deploying counter measures to blow up the torpedoes. They had their shields up so the lasers didn't penetrate. But they did not see the mines which were cloaked. When their ships hit the mines, they exploded and crippled both ships.

"Enemy ships have been disabled. Do you want us to take them out?'

"Fire a torpedo at each. Let's see how they deal with them when their mobility has been compromised."

Shooter sent out 2 more torpedoes. The enemies tried to blast them out of space but the aftershock destroyed them.

"How long till we reach GAP Command?'

"Sir, GAP Command on long range scanners. I am picking up debris. I think it is our fleet. It looks like we have been attacked,"

"On screen."

The team stopped in amazement as they viewed the wreckage. It would appear that Gap headquarters had been attacked in their absence. There were fighter ships destroyed and the debris was floating all through space. Alma navigated to GAP command and there was evidence the command center had been bombed. The question on everyone's mind was "how did this happen?"

"Space, land us. Alma, begin a scan of the area. Under no circumstances are you to interface with their system. Somehow you were bugged while we were in space so I don't want you to contract another virus. Ninja and Doc, check out the west sector of the station. Shooter and Space, take the east sector. Lady Hawk, you are with me to check out south sector. We all converge on north sector in 1 hour."

Ninja and Doc went through the west sector looking at the medical and research labs. There were no bodies around. As Doc examined the room he found that someone was working on experiments and had left them before they were finished. Ninja accessed the log and saw that they were examining something before they were taken away.

Space and Shooter checked out the east sector which housed the weapons along with air and ground fleets. Shooter saw that the weapons cache had never been opened. Space saw the fleets destroyed. It would appear that the fleet was hit before many of them were deployed.

Captain and Lady Hawk check out the officer and soldier quarters. They did not find the compliment of soldiers there they expected. They left the quarters and headed for the north sector which housed the command center.

The team all met at the north sector. Captain called for a progress report.

"We found evidence of the ships being bombed in the hangar. Weapons cache was not opened. Apparently someone or something hit this location hard and fast. They took out GAP from the inside. And we did not find any prisoners or bodies" said shooter.

"Appears there was some experiments going on in the lab but the results were non- conclusive" said Doc.

'We are in command center. Computer, this is Captain Matthew Lincoln. Protocol 8-76, voice verify."

"Captain Matthew Lincoln verified. What can I do for you Captain?"

"What happened to crew at GAP command?

'Crew has been taken captive."

"By whom?"

"DASHOOOR soldiers raided GAP command and took leaders captive. The other staff and personnel are being held in substation C holding cells."

'Computer, why did you not contact us and send out a priority alarm to orbiting space fleet?'

"Because that is not my programming. I am here to help DASHOOOR destroy GAP command and all of its Space Rangers."

"Who gave you that command?'

"You did sir."

"I did? Computer, I have not been here for over a month. Why do you think I gave you this command? And I do not have the authority to override the entire GAP command. I am only a Captain here. So why would you listen to me?"

"Captain, you are not making sense to me. Your comments are illogical."

"Ninja, Doc, Shooter, go free the prisoners. Bring officers back to the command center."

"Space, what is your analysis of this situation?"

"There is a new type of virus that has been able to infiltrate both Alma and GAP command. It seems able to change our programs and protocols. Without computer support I think the command center was a sitting duck. With no way to know the computer was compromised they didn't see it coming."

"What can we do to rectify this, Space? We still don't know their endgame" said Lady Hawk

"I can try to see if me and Alma can erase it."

"While you do that the rest of us will assess the situation here and begin to see who is left to help us get this base back on track. Keep me posted on your progress Space. Lady Hawk, see if you can get communication with Earth and tell them what has happened here. Doc, all survivors will be sent with sickbay so I want you to go and get ready to receive them. Shooter and Ninja, look for survivors and get them to sick bay. I will be in the command center. Stay in touch."

"Sir, are we sure we are free from hostiles?"

"Good observation Shooter. Before we begin can you perform a sweep of the base so we won't have any surprises?"

"Yes sir. Give me a few minutes before we all split up."

"While Shooter is at commander center with you performing scan Lady Hawk, Doc and I will go to sickbay and start getting it ready if that is okay with you."

"Great idea. Make it so Ninja."

Shooter scanned the base while Captain looked at the damage and began to put the command center back together. Suddenly he ran to Captain and yelled; 'Captain, sensors indicate we have 6 life forms here that are not GAP soldiers. I believe we have intruders still here."

"Lock down the base. Nothing leaves without my permission."

Captain hit his badge to communicate with the rest of the Rangers. "We have 6 intruders on this base that are still at large. Ninja, you all stay together and get sick bay operational. Be prepared for engagement. Space, me and Shooter are on our way to you. Stay put into we get there."

"Got you Cap" came from Ninja and Space.

"Shooter, let's get to Space now." With that they both ran to the hanger where Space was working with Alma. There were 4 intruders headed to Alma to commandeer her and leave the station.

Alma picked up the intruders on her sensors. "Space, we will have 4 intruders in 2 minutes."

"Lock your doors. We wait for Captain who is on his way. No one boards you. Contact Cap and tell him the situation."

"Captain, we are about to be attacked. 4 intruders headed to me with artillery."

"Almost there. Under no circumstances are you to let them board."

The intruders entered the hanger. They ran to Alma and saw she was not open. They tried to force her open but to no avail. Then they pulled their weapons. Just then Captain and Shooter hit the room.

'This is Captain Combat of the Space Rangers. You are all under arrest. Lay down your weapons and we will not hurt you. Resist and you will be subdued no matter the cost to you."

The intruders turned and began to shoot at Cap and Shooter. Shooter responded with rapid gun fire. Cap pulled out his Multi- gun. "Water pellets" and he shot all 4 intruders with pellets that exploded to water all over them. "Electric pellets" and he shot all 4 again. This time they were mildly electrocuted but not killed. However they were all disabled.

"Space we are outside. Enemy subdued. Shooter, tie all 4 and let's get them to interrogation."

"Alma, update."

"We are at 67% capacity. I am looking to replenish my power cells I also need to have my armaments restocked."

"Space, get Alma ready to leave this base as soon as possible. Shooter, I want you to tell me how they got in here in the first place. We need to close that hole."

"Yes sir. I am on it."

Doc and Lady Hawk got all the prisoners on the base freed. Their wounds were minor. It appears that Dashooor sent in enemy agents who took the base by surprise.

" Lieutenant Chambers, Captain Combat here. Can you tell us what happened to the base?"

"Sir we are attacked in open space as the fleet left to meet you at the Pluto station. The invaders took one of our ships and returned to base. They captured Head Command and took them away. The rest of us who were not killed were put into this detention."

"Lady Hawk I want you to use a sub sound space indicator to see if you can pick up any foreign signals on the ship. Doc I want you to scan the vessel for any non GAP personnel. Ninja do a sweep of the facility and make sure every way in is closed. Get with Shooter who is looking for any remaining intruders. I want this base secured before we leave here. "

"Yes Captain."

Captain, Lady Hawk and Lieutenant Chambers headed for the command center. Once there they began to put all systems back on line. Space had purged the computer system and all was back to normal.

Shooter and Ninja returned to the command center. "Sir we have secured the base. All entry points have been isolated and we have installed new security programs into our system."

Lady Hawk said "Sir a scan of the interior reveals we are not sending any new signals out."

"Lin said "Captain, I have scanned all remaining personnel on the base. We are back to all GAP personnel again."

"Great job team. Space, how long before Alma is space-worthy?"

"I will have her ready in 30 minutes."

"Good. Shooter, map a course for Dashooor world headquarters. We are going to attack the Citadel and take the fight to them."

"Yes sir. When do you want us to get ready for the attack?"

"In 30 minutes. Lieutenant, you are in command. Contact earth forces and get reinforcements up here pronto."

"I am contacting them now sir."

"Space Rangers, fill up your arsenal. We are headed into a dogfight."

Each member went to the weapons locker and replenished their supply. They knew they would need all the ammo they could get.

Chapter 10 – The Showdown

Lady Hawk approached Captain on the bridge. "Sir, can I speak to you a moment before we leave on this mission?"

"Of course. Walk with me to my quarters."

Once they got inside Lady Hawk fell into Captain's chest. She started to tear up.

"What is wrong? Don't answer that. I know what is wrong. And yes, I am scared too."

"Matt we may never see each other again. I don't want to lose you now. Not ever."

"Blair, I feel the same way. But we have to promise each other something. We both have to come out of this alive. We are too young with too much life ahead of us to go out like this. No matter what, these guys go down. Our team does not."

"I want so to believe that."

"Then do. There is nothing worse than believing you have failed before you even try. We have survived some hard times and we are still standing here today. Remember our encampment when we were all separated? You saved us honey. You were there and you saved me. Don't forget the stuff you are made of."

"That's why I love you. I reprogrammed Falco to offer great protection and more weapons. He will be watching both our backs down there."

"That's my girl." Then Matt pulled Blair close and kissed her passionately for what seemed like an eternity. When they released from their embrace they headed out of his cabin.

As they left Captain's quarters Alma came over the com for the Captain.

"Sir we will be entering DASHOOOR airspace in 2 hours and 14 minutes. What are your orders?"

"Let me know when we come out of hyper speed. I want us to come at them hot. Team, meet me in conference room in 5 minutes."

Everyone was assembled in conference room. "Alma, show me shuttle ships and motor bikes on screen."

"We have 3 high speed air bikes fully armed for close quarter combat. Ninja and Lady Hawk and I will oversee ground patrol. We will hit their ground forces hard and make our way to Citadel. Ninja, make sure we have full armament on board. We want to preserve our weapons for hand to hand combat. Space, you and Alma are to disable their air strip. Try to keep as many of these planes out of the sky as you can. Alma, how many ships and how many soldiers do you detect on the planet."

"I detect a total of 27 ships in airports and over 100 ground forces. They appear to be preparing for us as well sir. Long range scanners picking up airships orbiting the planet."

"Team we are here to disable and destroy this place. When we leave they must not be able to launch a counter strike Doc, tell us what you know about the duplicating of our badge technology."

"These badges operate like ours with one huge difference; they are the power source. Once you disable the badge, they lose their powers."

"That will be our strategy then. We will take their badges away from them. Doc, are they all independently powered or is there a main source providing power to all badges?"

"Good question Cap. Lady Hawk, I need you and Alma to assist me in determining power source. If we can identify how they power these things maybe we can disable them."

"That would turn the battle in our favor. If we are evenly matched on the ground they outnumber us 5 to 1. Not great odds."

Doc and Lady Hawk went to science lab. With Alma's help they were able to isolate the power source. They ran to tell Captain.

'Sir, the power source. The planet is the power source. These badges are all connected to a main battery which appears to be about 100 meters below ground. If we can disable that battery we can knock these badges out. I am noticing a power surge building up even as we speak. I think they are powering up their badges in anticipation of us.

"Great news. That means we have to attack in the air and on the ground and under the ground. Ninja, once we start the ground attack you and Lady Hawk will keep the fight going. I will head to the battery and disconnect it."

"It will be heavily guarded Cap. I don't think you can reach it alone. You need to take one of us with you."

"Shooter and Doc will rendezvous with you on the surface. Once they get there send me some back up. I will try to hold them off until you arrive."

'Yes sir. All vehicles loaded and ready to go. We are ready to launch our attack."

"Listen I know we have not used our badges continually and now we may have to. Try to avoid it as long as you can. But do what you have to in order to survive. That is an order."

"Sir we are approaching enemy space. There are 5 battle cruisers waiting for us to exit hyper space."

"Alma, what is their armament? Can we take them?"

"They carry a compliment of torpedoes and lasers. Our shields can withstand the armament unless they all focus at one area. Then they will penetrate my shields and breach my hull."

'Space, make sure that does not happen. You have to get us past this initial defensive wall. Shuttle can't evade their battle ships."

"I got you sir. I will get us to the surface air space."

"Prepare to engage. Leaving hyper space in 3 minutes."

"Red alert. All hands on deck. Once in surface air we will depart to our mission. Shooter, take them out."

The ship maneuvered out of hyperspace and was hit by enemy lasers. Alma's shields were holding.

"Now Shooter. Light them up."

Shooter had already programmed certain offensive moves into his console so that all he had to do was target and execute.

"Alpha phase 1 – attack" and a barrage of laser blasts hit the waiting enemy ships. "Delta phase 1 – attack" and torpedoes were launched at each ship. Every one was a direct hit.

"Alpha phase 1 – attack" and lasers shot out again at the exact point the torpedoes hit. 3 ships exploded and two more were sent back into orbit headed to the planet on fire.

"Our job is to get this done quickly so we can attack from the rear and back up Cap and the team. Set targets and fire at will."

As Doc flew the shuttle Shooter delivered all torpedoes as needed. The mining company and all the open mines were destroyed. The men on the ground scrambled to get out of the way of the exploding equipment and take cover. Doc landed the ship and he and Shooter left the shuttle headed for the Citadel. As the guards came running out to defend the Citadel Doc and Shooter began to fire upon them. They made a way to the entrance of the Citadel and went inside.

"Okay, let the games begin." With that Shooter and Doc began placing smart bombs on the pillars inside looking to blow the Citadel from its foundations. They took out guards along the way.

Alma and Space launched an aerial attack. She sprayed the area with an array of bullet fire. Then she fired torpedoes at each ship attacking their shields. The lasers were pointed at each weak spot and penetrated the areas of weakness. Then Alma sent 4 torpedoes to the air base hub and destroyed the landing strip. Space shot up all planes that were docked until they exploded.

"Alma, shoot a torpedo at the Citadel now."

As the torpedo left the weapons bay the remaining air ships scrambled to catch the torpedo before it could hit the Citadel. This allowed Space to completely decimate the air field. The air ships stopped the torpedo and looked back to see the air strip destroyed. They all turned and began firing at Alma.

"Get us out of here. We need some room to take care of them." With that Alma rocketed off to space. The enemy ships followed.

Captain Combat, Lady Hawk and Ninja roared across the plains headed to the Citadel. There were ground crews assembled and they opened fire. Captain and his team converted their cycles to air mode and began shooting at will. A deflector shield came up and the fight was on.

The three of them went straight for the towers near the door of the Citadel. There was a barrage of shots as the soldiers were hit. Cap and his team hit their badges and their body armor came into play.

"I will get word to the Captain. You lock in on their life signatures. Space, Shooter here with Doc. We are picking up human life forms 3 levels down. Can Alma scan and tell us what we are dealing with."

"Yes. Alma, deep scan below the surface and find those human life forms."

"Sure Space. I have them on my sensors now. I am running a scan. Space, we have a problem."

"What is the problem?"

"I have identified the life forms. Their DNA match 6 members of GAP."

"Those hostages are family members of GAP?'

"Yes."

"Do you know whose family we are talking about? Commander? Colonels?"

"No. They are family members Space Rangers Alpha Team."

Shooter and Doc stopped walking. They looked at each other.

'Alma, they are holding our families?"

"Yes. And some are injured although all seem to be alive."

"How many?"

"I count 17 total. Each of you has at least one family member being held captive. They are 20 meters ahead of you behind a steel door."

"Relay info to rest of team. We are going to see about getting them out."

'Shooter, wait for Cap's orders. We have a mission to complete."

"I am going to get our family. Tell Captain the situation. Could use back up when available."

"I will inform them and help is on the way."

Captain, Lady Hawk and Ninja were busy taking care of DASHOOOR troops. The battle was pretty evenly matches. What they lacked in armament they more than made up for it in manpower.

"Cap, space here. Shooter and Doc discovered humans on the 3 level. Going in to rescue."

"Humans? Prisoners here? How did DASHOOOR capture humans? Have you identified who they are?"

"Yes sir. They are our families from earth."

"WHAT? They have our families here? Here?"

"Yes. Shooter and Doc say some are injured and they are close so going to rescue."

"Space, you get our families out of there. Give them back up. We will take care of up here."

"But Cap, you all need help. There won't be support coming from the back for you."

"You have your orders Space. You go help them. Alma, you will be our back up, Get around behind this army and open fire at will."

"Understood sir."

Space armed himself for battle and grabbed some extra medical emergency kits. Then he grabbed a gyro cycle and left the ship headed for the back entrance of the Citadel. He saw reinforcements headed there to take on Shooter and Doc. He began shooting them as he drove past them on his cycle.

Alma positioned herself above the fight and maneuvered until she was behind the front ground attack. Then she began to open fire at will and mowed down dozens of DASHOOOOR soldiers at a time.

Captain observed bunkers opening and tanks revving up to come out and attack Alma. "Ninja ,Lady Hawk; take out those tanks. I got the foot soldiers."

Ninja headed to bunkers closest to him. He jumped inside and took out the four soldiers looking to get into the tank. Then he opened the door and dropped a grenade into the tank. It blew up for the inside and was useless.

Lady Hawk went to the other bunker. She found three soldiers and engaged them in battle before taking them out. When the soldier inside the tank opened a window to try to get a shot at her Falco flew at the window and dropped a grenade inside. As the soldier turned Falco grabbed the gun and pulled it through the window. The grenade went off and the tank was destroyed internally.

Ninja and Lady Hawk left the bunkers and ran looking for Captain.

As he approached the ground soldiers Captain put his rifle on his shoulder. Then he drew his Multi- gun and began shooting at the soldiers.

"Missile projectiles." The grenades shot into the air and exploded on contact near the soldiers sending them flying.

"Rapid fire sidearm." He shot a succession of bullets at the soldiers taking those who were still standing out.

"Gas projectiles." Captain shot into the air and green smoke filled the area. The soldiers began coughing and passing out as the gas was poisonous to them. By now Ninja and Lady Hawk joined Captain.

"Alma, sweep the area and keep us posted. We will be leaving in a hurry so be ready for emergency evacuation."

"Yes Captain."

He motioned to Lady Hawk and Ninja to join him as he entered the Citadel. They had taken out over 100 soldiers and were now inside.

Meanwhile Shooter and Doc were at the entrance to the holding cells. Soldiers came around the corner and Shooter began firing. "Doc, get that door open now. Any means necessary."

"Gotcha". With that Doc placed the explosives on four corners of the doors and told the prisoners to get back. Then he ran from door and ignited the explosives. The door flew about 10 feet from the cell. Doc and Shooter went inside. They found all their family members in there. A few were injured. Doc tried to help them as quickly as he could. Shooter watched the door.

More soldiers came running down and caught Shooter in a crossfire as they hit the prison cells from two directions. Shooter was doing his best but he was pinned.

'If we don't get help soon we are going to be stuck here."

"No, you will not." And Shooter and Doc turned to see Space ride down one stairwell on the gyro cycle shooting all soldiers in his way. Then Shooter turned 100% of his attention to the remaining stairwell. Space threw a grenade inside and blew all the soldiers away.

"Great to see you. Knew you would get here eventually."

"Yeah, yeah, yeah. What do we have going on in there?"

"Plenty of scared people. But we have to go. We don't have time because reinforcements are coming."

"Doc, let's move them out. Me and Space will provide cover. Got to go now."

"Here we come. It's 17 total down here. Prepare Alma for injuries."

"Captain, Shooter here. Captives in our custody. Headed outside. We need a pickup.'

"Get to pick up point and have Alma take the people. Space, you and Doc get on board and get ready to pick us up. Shooter, I need you to get to me as soon as you have them on board."

"Understood."

The group ran outside and to the area where the pick up was to occur. Alma was waiting for them. They all hurried aboard.

"Listen I am going to provide Captain and the team a way out. Space, keep me on your trackers and give me some aerial support if I need it."

"Got you."

Meanwhile Captain, Lady Hawk and Ninja continued to make their way to the Citadel capital. The soldiers sent to engage them were no match for the firepower the Space Rangers wielded. As they entered into the entrance of the Capital they saw a form that was large. It was over 9 feet tall with a club in his hand. He was the one they called "Goliath".

"Pity humans. Now you have angered the Master and you shall pay with your lives."

"And who is your Master?" yelled Captain.

"I shall take your head to meet him personally."

"Ninja, Lady Hawk assume flank positions. Watch out for that club of his."

"The three of you are no match for me. I was created to withstand all of the weapons you carry. So how you are you going to stop me?"

"Maybe by talking less and striking more." With that Ninja began blasting Goliath with pellets from his staff. Goliath winced but did not go down.

Lady Hawk loaded her rifle. She began firing on Goliath left side. He looked at her and you could tell he was feeling the bullets but refused to go down.

Then Captain loaded his rifle with an explosive pellet and hit Goliath point blank. The blast knocked Goliath back 10 feet. But he did not stay lying down. He got up and dusted off his chest.

"So the little people think they can stop me. Well, as you can see, I can take what you dish out. But can you?" With that Goliath shot out a shock wave from his massive club. The blasted leveled all the Space Rangers. Goliath laughed.

"I c an take you all. Just watch." With that Goliath hit his badge But instead of activating nothing happened. He looked puzzled.

"What's the matter bighead? You having a little trouble there?"

"No, no no. This cannot be. I was bred to destroy you. I have the power. You cannot stop me."

"Yes we can and we will." With that Ninja hit Goliath with a barrage of explosive ninja stars. The blast surprised Goliath. For the first time he felt pain.

"Lady Hawk, you and Falco take the left side. Ninja, the right. I got him head on."

Lady Hawk took out her electric whip and began hitting Goliath with electrical shocks. Falco hit his head with laser beams aimed at his helmet. As the helmet heated up Goliath took it off and threw it at Falco. He dodged the helmet and hit it again with his laser eyes. Lady Hawk sent an electrical blast to Goliath face. He winced in pain.

Ninja went and engaged Goliath in hand to hand combat. Although he did not have the benefits of using his badge Goliath was still stronger than any normal man.

Cap took out his Multi gun and pointed at the ground beneath Goliath. "Heat Ray." The ground began to melt under Goliath. He started to sink. He wanted to escape but could not pull his feet up. "Liquid Nitrogen" yelled Cap and he froze the ground under Goliath. Now he had sunk with his knees under the ground frozen. Then Ninja began to kick and hit Goliath. He was blocking his blows. He picked up his club to fight and Lady Hawk wrapped his club in her whip. She electrified it and the shock hit Goliath.

"Liquid Metal" and the gun shot out waves of liquid metal that hit Goliath and began to harden. In time his whole upper body was covered in the metal. He tried frantically to move.

"Water pellets" with each blast the pellets exploded into water and covered Goliath. Then Lady Hawk wrapped him with her electric whip and turned on the power. Goliath fried like a piece of bacon.

As the team left the Citadel smoke was everywhere. It was in a state of destruction. Their mission was a success.

But there were many unanswered questions. How did they get set up when Philips badge was taken? And who attacked GAP command?

The Rangers realized their work had just begun.

If you enjoyed this space adventure, look for our next book in this series.

Space Rangers are here to stay.

Dr Jeff Davis
Author

www.ingramcontent.com/pod-product-compliance
Lightning Source LLC
Chambersburg PA
CBHW080828180526
45168CB00006B/2614